SpringerBriefs in Electrical and Computer Engineering

Series Editors:

Woon-Seng Gan, School of Electrical and Electronic Engineering,
Nanyang Technological University, Singapore, Singapore
C.-C. Jay Kuo, University of Southern California, Los Angeles, CA, USA
Thomas Fang Zheng, Research Institute of Information Technology,
Tsinghua University, Beijing, China
Mauro Barni, Department of Information Engineering and Mathematics,
University of Siena, Siena, Italy

T0171976

SpringerBriefs present concise summaries of cutting-edge research and practical applications across a wide spectrum of fields. Featuring compact volumes of 50 to 125 pages, the series covers a range of content from professional to academic. Typical topics might include: timely report of state-of-the art analytical techniques, a bridge between new research results, as published in journal articles, and a contextual literature review, a snapshot of a hot or emerging topic, an in-depth case study or clinical example and a presentation of core concepts that students must understand in order to make independent contributions.

More information about this series at http://www.springer.com/series/10059

Iraj Sadegh Amiri • Masih Ghasemi

Design and Development of Optical Dispersion Characterization Systems

 Springer

Iraj Sadegh Amiri
Computational Optics Research Group
Advanced Institute of Materials Science
Ton Duc Thang University
Ho Chi Minh City, Vietnam

Faculty of Applied Sciences
Ton Duc Thang University
Ho Chi Minh City, Vietnam

Masih Ghasemi
Institute of Microengineering
and Nanoelectronics
Universiti Kebangsaan Malaysia
Selangor, Malaysia

ISSN 2191-8112   ISSN 2191-8120   (electronic)
SpringerBriefs in Electrical and Computer Engineering
ISBN 978-3-030-10584-6   ISBN 978-3-030-10585-3   (eBook)
https://doi.org/10.1007/978-3-030-10585-3

Library of Congress Control Number: 2018967262

This Springer imprint is published by the registered company Springer Nature Switzerland AG
The registered company address is: Gewerbestrasse 11, 6330 Cham, Switzerland

Abstract

This book presents the implementation of automated measuring system for measuring chromatic dispersion by using modulation phase shift method over long haul of optical single mode fibre. All forms of dispersion degenerate the modulation-phase relationships of light wave signals, decreasing information-carrying capacity through pulse-broadening in digital networks and distortion in analog system. Hence, designing system for measuring chromatic dispersion might be essential for estimation the performance of optical transceiver systems. The scheme for measuring chromatic dispersion is adopted in conjunction with tunable laser (TLS) to provide the optical power at required wavelength and digital oscilloscope (DOSC) for measuring phase difference between microwave signal from transmitter and microwave signal at receiver. The working principles of the systems are comprehensively elucidated in this book. Additionally, the system components of the designs are identified, and the systems are thoroughly characterized.

Contents

Chapter 1
Concepts and Fundamental Theories of Optical Fibre Dispersions

1.1 Motivations

In optical communication, there are different types of distortion that cause the received optical pulse shape to deform in irregular manner. This distortion which is mainly due to dispersion could degrade the phase of light wave signal and reduce the capacity in digital networks. Multimode fibre and single-mode fibre inherently lead to broadening of pulse, which is caused by three basic forms of fibre dispersion, namely, intermodal, chromatic and polarization-mode dispersions, respectively.

Data rate is limited by intermodal dispersion in multimode fibre because the travelling distances are slightly different for propagated signal modes and in chromatic dispersion is also limited because in high-speed communication, each wavelength component of the signal travels at different speed of propagation. In polarization-mode dispersion, the orthogonal-polarized components of optical pulse have different propagation velocity in single-mode fibre.

All the above facts somehow related with broadening of optical pulse in fibre medium and deterrent factors in designing optical communication systems. Measuring dispersion to characterize the optical system response in terms of wavelength, frequency and distance is the interest motivation for the subject of this book.

1.2 Objectives

In any optical systems, both optical device response itself and combination of optical devices will have different responses. For example, the electro-optic modulator with or without employing polarization controller gives different responses to the

I. S. Amiri, M. Ghasemi, *Design and Development of Optical Dispersion Characterization Systems*, SpringerBriefs in Electrical and Computer Engineering, https://doi.org/10.1007/978-3-030-10585-3_1

1

optical sources such as tuneable laser. Therefore, the optical devices need to be characterized based on their configuration and applications. The following objectives are the major consideration for the current work:

- To design and develop algorithms for characterization and controlling optical apparatus
- To study and design automated characterization system for electro-optic modulator with application of driving microwaves
- To design an algorithm for automated dispersion measurement for different length of fibres
- To study and analyse the effects of changing critical parameters such as frequency, wavelength and length of fibre on dispersion

1.3   Major Contributions of the Book

The aggregate challenges on this work are aim at increasing accuracy and speed of measurement, achieved via convenient control of optical apparatus via computer. The following contributions are listed as major objectives:

- Statistical analysis to find the best voltage biasing value point of electro-optic modulator
- Analysis of average phase shift of propagating light pulse versus driving microwave frequency for optical modulator
- Analysis of phase shift of propagating light pulse versus wavelength
- Effect of changing biasing point of electro-optic modulator on measured phase shift of light pulse propagation

1.4   Organization of the Book

The current work begins with a brief introduction, followed by a summary of objectives and contributions in this chapter. Chapter 2 explains any relevant basics and fundamental theories related with optical fibre and all kinds of dispersions. Chapter 3 describes all the employed facilities, instruments and measuring devices. Required conditions for normal operation of relevant devices are also discussed in this chapter. Methods of measuring of dispersion are also summarized for more comparisons. Chapter 4 proposes some schemes for examining devices and also algorithms to be applied to the programmable devices in order to reach the objectives listed in Sect. 1.3. Chapter 5 shows the results and achievements about characterizations of device in terms of DC bias voltage point, wavelength, frequency, dispersion and length of fibre.

Chapter 2
Single-Mode Optical Fibre Dispersions and the Physics Phenomenon Involved

2.1 Overview

This chapter reviews the literature concerning types of dispersion caused by a single-mode optical fibre. As a starting point, Sect. 2.2.1 reviews the single-mode fibre characteristics in one glance. Section 2.2.2 lays out the theory on group-velocity dispersion (GVD). Section 2.2.3 subsequently shows how polarization of light in a waveguide medium leads to dispersion. Section 2.2.4 explains the basics of dispersion on a waveguide. Section 2.2.5 illustrates that a material can also take dispersion into account. Finally, all the information given in this chapter is summarized in Sect. 2.3.

2.2 Fundamentals of Optical Dispersion

Many telecommunication companies and internet service providers these days are facing an unceasing demand for accessibility to the internet or networks. Installing new long-haul links to carry the growing traffic means enormous cost, so another way to meet this demand is to improve the capacity of the available link. Here, the proposed question is which type of link has the ability to enhance the capacity.

The optical fibre, because of its significant advantages over other mediums, such as wire or coaxial cable, has been used as an excellent transmission medium by the telecommunication industry since 1974. In addition, the remarkably low attenuation, the small size, the high data transfer rate, not being interposable with electromagnetic fields, and the wide bandwidth, among others, are taken into account for an optical fibre.

This part focuses on the physical characteristics of the fibre and on the effect of the fibre as a medium on the optical pulse as it propagates through the fibre. The

I. S. Amiri, M. Ghasemi, *Design and Development of Optical Dispersion Characterization Systems*, SpringerBriefs in Electrical and Computer Engineering, https://doi.org/10.1007/978-3-030-10585-3_2

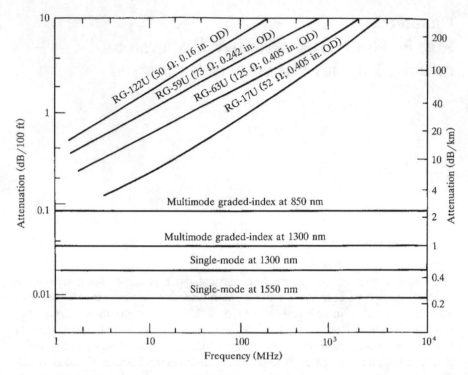

Fig. 2.1 By increasing frequency, the attenuation curve for single-mode and multimode fibres is completely flat compared with a coaxial cable

historical development of a fibre shows that many innovations have been made to configure fibre as an acceptable medium for transferring optical data, but it is still not considered as a perfect medium and has its own pros and cons.

In generally, two groups of fibres, single-mode and multimode, are used as the means for different purposes and applications. Because the single-mode fibre is chosen for all the experiments in this book, referring to retaining accuracy of the injected optical pulse in the long haul and providing higher bandwidth compared with multimode fibres and also coaxial cable, such as observed in Fig. 2.1, we study all the fundamental dispersions on single-mode fibres in this chapter. Because frequency dependency of group velocity in the single-mode fibre is caused by intramodal dispersion or, in other words, fibre dispersion, in the next section the theory of the fundamentals of dispersion on the single-mode fibre is studied first.

2.2.1　Propagation Mode in Single-Mode Fibres

An optical fibre alone can be specified as a dielectric environment (waveguide) with performance in the optical frequency range. Light, which is classified as a sort of electromagnetic field, is guided through the fibre in such a way that its properties are

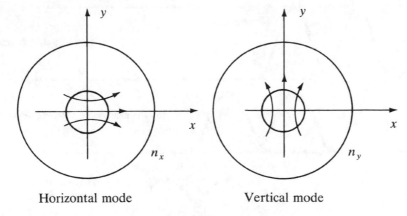

Horizontal mode   Vertical mode

Fig. 2.2   Polarizations of fundamental mode HE_{11} in single mode

entirely dependent on the physical structure of the waveguide. The purpose of guided light is referred to those guided electromagnetic fields that propagate along the fibre, also called waveguide modes. Guide modes are the form of electric and magnetic field distributions that occur in a cyclic manner [1, 2].

The allowable core diameter for a single-mode fibre is usually between 8 µm and 12 µm, allowing only a small number of wavelengths to pass through the fibre. In the practical design of a single-mode fibre [3], the core cladding index difference varies between 0.2% and 1.0%. Because geometric parameters of light distribution can be considered as an important factor rather than other factors such as core diameter and numerical aperture, this parameter is used to analyze and anticipate the efficiency of a single-mode fibre, and the basic mode in a single-mode fibre is mode-field diameter (MFD) [4, 5]. Figure 2.2 shows two other regular orthogonal modes in a single-mode fibre, which are called the horizontal (H) and the vertical (V) polarization modes. Each of these two modes can be used for establishing the primary HE_{11} mode [6, 7].

Almost any defect that leads to an asymmetrical shape for the circular shape of a fibre causes a phase difference between the propagation of the two modes. As a result, the difference between the effective refractive indices of the two modes is called the fibre birefringence.

2.2.2   Chromatic Dispersion

In simple words, chromatic dispersion (CD) is caused by a slight change in the refractive index of a single-mode fibre when the wavelength is altered. At some wavelengths it can be seen that the group velocity of a light wave has different velocities and is traveling faster than other groups. For this reason, CD, which means pulse broadening, is also called group-velocity dispersion (GVD), and related to this phenomenon, optical pulses comprise a range of wavelengths, and during

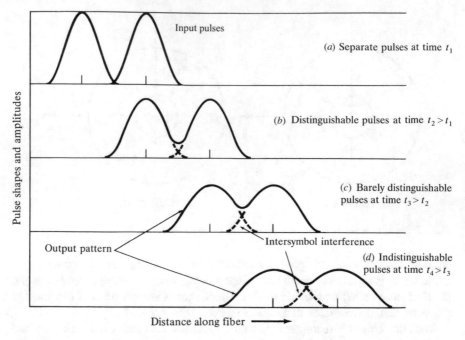

Fig. 2.3 Broadening, attenuation and overlapping of two pulses as they pass through a fibre

travelling inside a medium their phase is changed. In other words, an optical pulse consists of some components such as wavelength and polarization state, each travelling at a different speed inside the fibre. Figure 2.3 shows two nearby pulses broadening and attenuating. By persisting in time, the two adjacent pulses become wider and wider so that they are not distinguishable and overlap each other. Usually, measuring the bandwidth that lets the information be carried on the waveguide is called bandwidth-product in MHz km.

Optical pulses are received at the end of a fibre at different times, which could be interpreted as delay. For a certain length of fibre L, the delay for a specific spectral component at frequency ω, the time delay when it arrived at the end of the fibre is $T = L/V_g$, where V_g is group velocity. To calculate group velocity, the formula is defined as follows [1]:

$$V_g = \left(\frac{\mathrm{d}\beta}{\mathrm{d}\omega} \right)^{-1} \tag{2.1}$$

Here β_1 is the propagation constant along the fibre and V_g is the velocity of the optical pulse energy along the fibre. β is also called β_0 and is the first component of the $\beta_L(w)$ expanded Taylor series. The first three components of the Taylor series provide a good estimation for $\beta_L(w)$, and hence can be used for any fiber. $\beta_2 = \dfrac{\mathrm{d}^2\beta}{\mathrm{d}w^2}$ and $\beta_3 = \dfrac{\mathrm{d}\beta^3}{\mathrm{d}w^3}$ are called the second- and third-order dispersion parameters, respectively, and are the reasons for pulse broadening in optical fibres. Dispersion in terms of wavelength is shown in Eq. (2.2).

$$D = \frac{\mathrm{d}}{\mathrm{d}\lambda}\left(\frac{1}{v_g}\right) = -\frac{2\pi c}{\lambda^2}\beta_2 \tag{2.2}$$

The unit of dispersion is ps/km nm. It is clear from the foregoing formula that dispersion is a nonlinear function of wavelength and for different wavelengths we have different values for dispersion; however, for zero-dispersion wavelength, this nonlinearity is changed to an approximately linear function [8, 9].

$$D = S\left(\lambda - \lambda_{ZD}\right) \tag{2.3}$$

Here λ_{ZD} is related to the fibre and S is the dispersion slope [10], and this parameter is also a function of λ, written as follows:

$$S = \left(\frac{2\pi c}{\lambda^2}\right)^2 \beta_3 \tag{2.4}$$

2.2.3   Polarization Mode Dispersion

As discussed in Sect. 2.1, fibre imperfections cause the circular shape to be nonhomogeneous, and its physical structure changes along the fibre randomly. In addition, these physical imperfections can be induced by the environment. Practically, the process of changing state of polarization (SOP) does not concern optical devices. The information also is not coded using polarization, and in the detection scenario the photodetector just captures the total incident power, which is not included in the SOP standalone. In spite of all these factors just mentioned, there is one principal mode leading to broadening of an optical pulse with variable magnitude, named the polarization mode state (PMD). Because the source of producing PMD relies on physical factors from the environment, it is not preventable. The PMD for an optical system at a lower data rate does not cause many problems, but for a higher data rate in the range of gigabits per second (Gb/s), where the optical pulses become narrower, it can be the origin of destructive error propagation. Related to birefringent matter, there are two definitions:

1. Fibre with constant birefringence:

In Sect. 2.1, it is described that a single-mode fibre has two modes of polarization that are perpendicular to each other. If we consider our fibre is ideally standard and not affected by physical problems, the two modes propagate at the beginning of the fibre but along the way one of them moves slower than the other. If the difference between the two modes is called β_p, it can be expressed as the index difference between two basic axes and written as follows:

$$\beta_p = 2\pi / \lambda \left(n_x - n_y\right) \tag{2.5}$$

One of the two axes must have a faster propagation of mode than the other because the SOP of each axis is not stimulated simultaneously and also each one has a different phase and group velocity. As a result, the original pulse breaks into two pulses that are orthogonally polarized. The delay between two consecutive pulses can be estimated in Eq. (2.6) and is called differential group delay (DGD).

$$\Delta\tau = \left| \frac{L}{v_{gx}} - \frac{L}{v_{gy}} \right| = L\left(\beta_{1x} - \beta_{1y}\right) \tag{2.6}$$

2. Fibre with random birefringence:

All the facts about constant birefringence are for a short length of fibre in which the fibre is affected just by the common environmental situation. However, in the long haul, for a fibre more than 1000 kilometers (km) many environmental parameters are changed. The facts about the long haul return to the nature of birefringence, its tendency varying in short sections of length (might be about 10 m). So, for the long haul the random changes in birefringence caused pulse distortion along the fibre. For different fibre sections, in spite of similar lengths for each one and for different sections, birefringence is different in terms of magnitude and the orientation of the principal axes. This random manner causes the final prediction of $\Delta\tau$ to become impossible. The final result shows the optical pulse at the end of the fibre is distorted and in some cases its time location is shifted to another location. Such distortion, which is caused by the PMD, has a negative effect on the reliability of the light wave system and needs to be studied to remove its effect.

2.2.4 Waveguide Dispersion

Waveguide dispersion occurs when only a portion of the light travels in the core and the remainder of it travels in the cladding. Cladding, because of the different refractive index and other characteristics, causes the confined portion of light inside to travel faster than in the core. This difference is called waveguide dispersion and leads to pulse broadening in terms of different wavelengths. For different types of fibre, waveguide dispersion gives different responses; hence, it is important to study the parameters associated with this dispersion.

$$\left|D_{wg}\left(\lambda\right)\right| = \frac{n_2 \Delta}{c} \cdot \frac{1}{\lambda}\left[V\frac{d^2\left(Vb\right)}{dV^2}\right] \tag{2.7}$$

Here, n_2 is cladding refractive index, Δ is index difference, and V is another form of definition for the propagation constant, called the normalized frequency Eq. (2.8). The second order of group delay $\dfrac{d^2(Vb)}{dV^2}$ is represented in the form of derivation ratio. In Eq. (2.8), the parameter shown is core diameter.

$$V = \mathrm{kan}_2 \sqrt{2\Delta} \qquad\qquad (2.8)$$

2.2.5 *Material Dispersion*

Material dispersion occurs when the refractive index of a fibre is as a function of wavelength. Group-velocity density is related to refractive index. Concerning this fact, somehow it becomes a function of wavelength. Equation (2.9) shows how this type of dispersion is related to wavelength and refractive index.

$$\left| D_{\mathrm{mat}}(\lambda) \right| = \left| \lambda \cdot \frac{d^2 n}{d\lambda^2} \right| \qquad\qquad (2.9)$$

Usually the summation of material dispersion and waveguide dispersion is known as dispersion and provides a good estimation instead of using the direct calculation formula, which brings in Eq. (2.2). Thus, , chromatic dispersion can be rewritten as follows:

$$D = D_{\mathrm{mat}}(\lambda) + D_{\mathrm{wg}}(\lambda) \qquad\qquad (2.10)$$

The following Fig. 2.4, shows the waveguide dispersion and material dispersion trends for different values of wavelength. From this figure it is also clear that, for lower wavelengths, waveguide dispersion effects are almost considerably higher compared with material dispersion; however, for higher wavelengths, material dispersion offers a substantial upward trend.

2.3 Summary

The current discussion for single-mode optical fibres originates from the general dispersion group called intermodal. Parameters such as wavelength and fibre length are considered as critical. Their changes cause the received amplitude, phase, and pulse duration of the optical pulse to vary. All factors concerning the foregoing discussion are diminished performance of the optical system after modulation.

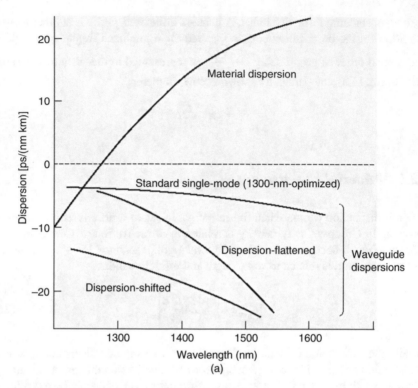

Fig. 2.4 Intermodal dispersions against changing wavelength

Additionally, these troubles cause the perfect shape of the received optical pulse to be limited by the optical fibre, modulator, and optical receiver properties between transmitter and receiver.

References

1. K. Wakita, I. Kotaka, H. Asai, IEEE Photon. Technol. Lett. **4**, 29 (1992)
2. E.A.J. Marcatili, Objectives of early fibres: evolution of fibre types, in *Optical Fibre Telecommunication*, ed. by S. E. Miller, A. G. Chynoweth, (Academic, New York, 1979)
3. D. Marcuse, D. Gloge, E.A.J. Marcatili, Guiding properties of fibers, in *Optical Fibre Telecommunication*, ed. by S. E. Miller, A. G. Chynoweth, (Academic, New York, 1979)
4. M. Artiglia, G. Coppa, P. Di Vita, M. Potenza, A. Sharma, Mode field diameter measurements in single-mode optical fibres. J. Ligthwave Tech. **7**, 1139–1152 (1989)
5. ITU-T Recommendation, *G.650 Definition and Test Methods for the Relevant Parameter of Single-Mode Fibres* (ITU Publications, Geneva, 1933)
6. L.B. Jeunhomme, *Signgle-Mode Fiber Optics*, 2nd edn. (Dekker, New York, 1989)
7. I.P. Kaminow, Polarization in optical fibers. IEEE J. Quantum Electron. **QE-17**, 15–22 (1981)
8. Y.W. Li, C.D. Hussey, T.A. Birks, J. Lightwave Technol. **11**, 1812 (1993)
9. J. Kani, M. Jinno, T. Sakamoto, S. Asiawa, M. Fukui, K. Hattori, K. Oguchi, J. Lightwave Technol.
10. Y. Mochida, N. Yamaguchi, G. Ishakawa, J. Lightwave Technol. **20**, 2272 (2002)

Chapter 3
Study of Optical Fibre Dispersion and Measuring Methods

3.1 Overview

Generally, the design of apparatus configuration plays fundamental and deterministic role over acquired outcome. For any suggested model, there would be many devices available in the laboratory that meet our required expectation on the quantity of proposed variables or parameters. However, some of the devices do not meet the desired reliability, stability or accuracy quality. For the intent of this book, in addition of the above qualities, the response time of measuring device is quite a significant factor. In the following sections, important features and characteristics of active or passive elements will be investigated to exploit them in implementing the design for characterizing dispersion of the optical field.

3.2 Dispersion Measurement Methods

This section first starts to study the available methods on measuring chromatic dispersion before going on more details about experimental device to find out the compatibility of device and method of measurement. The common techniques for measuring dispersion can be classified into some groups:

1. Time of flight (TOF)
2. Modulation phase shift (MPS)
3. Temporal interferometric
4. Spectral interferometric

© The Author(s), under exclusive license to Springer Nature Switzerland AG 2019
I. S. Amiri, M. Ghasemi, *Design and Development of Optical Dispersion Characterization Systems*, SpringerBriefs in Electrical and Computer Engineering, https://doi.org/10.1007/978-3-030-10585-3_3

3.2.1   Time of Flight

In this technique, two mechanisms can be employed to predict second-order dispersion parameter. First is by measuring the span of variable time delay between consecutive pulses at different wavelength, and the second is by measuring amount of widening of the pulse at the beginning and end of fibre [1]. The formula to determine dispersion is expressed in Eq. (3.1) [2]. The parameters Δt, $\Delta \lambda$, λ_0 and $D(\lambda_0)$ are effective group delay, pulse wavelength bandwidth, centre of wavelength and chromatic dispersion coefficient for above equation.

$$D(\lambda_0) = \frac{\Delta t}{L \Delta \lambda (\lambda_0)} \tag{3.1}$$

The serious problem associated with this technique is about length of fibre. For the short length of fibre, the measured dispersion could not be accurate because the given delay is not enough to distinguish the time difference between pulses. However, for long length (above several km), it is possible to measure dispersion on the order of 1 ps/nm.

3.2.2   Modulation Phase Shift

The MPS technique exploits the variation of phase of radio-frequency (RF) modulation envelope, while the wavelength is changed. The component for providing variable wavelength laser is usually tuneable laser with resolution bandwidth in the range of 1 nm. For implementing RF modulation, an external modulator is employed. The RF envelope and laser applied to external modulator and at the receiver amplify before detection by optical detector. If the RF envelope is considered as baseline, then by comparing the phase and amplitude of recovered data with detected optical envelope, it is possible to measure group delay over a wavelength $\Delta \lambda$. The following formula is used for calculation of group delay from obtained phase data:

$$\Delta T_g = \frac{\Delta \Phi}{360} \times \frac{1}{f_m} \tag{3.2}$$

where the first parameter $\Delta \Phi$ represents RF phase shift and the second parameter shows frequency of applied sine wave. To calculate chromatic dispersion, usually the ratio of group delay to the difference wavelength is used.

$$D = \frac{\Delta T_g}{\Delta \lambda} \tag{3.3}$$

Practically, to achieve higher precision in the range of 1 ps/nm-km, the length of fibre at this method must be tens of metres [2]. The other significant limitation factor is about accuracy for different length of fibres [2]. Measured dispersion for long length fibre could be constant with acceptable tolerance value, but for short section of fibre, it would not be acceptable. This is due to the fact that the uniformity for the long length of fibre is considerably changed compared with short length of fibre.

3.2.3 Temporal Interferometric

This technique can be implemented by using a wideband source and the variable path. For measuring dispersion on short fixed length of fibre, the variable air path is employed to produce a temporal interferogram (Fig. 3.1) between it and fixed length of fibre which is under the test. The wideband source is utilized by two temporal interferometry arms [3–9]. One arm is known as reference and the other one is unknown test past. This arm is moving at constant speed, and it is possible to plot the output intensity as function of time. By applying Fourier transform, it is possible to extract the pattern of phase spectral information that is usable for calculating dispersion. The second derivation of this phase spectral gives the dispersion information indirectly.

3.2.4 Spectral Interferometric

This technique is also used for measuring dispersion in short length of fibre. Here the spectral interferogram can be extracted from wavelength domain. There is no need to move one arm because both arms are fixed, so this method offers more

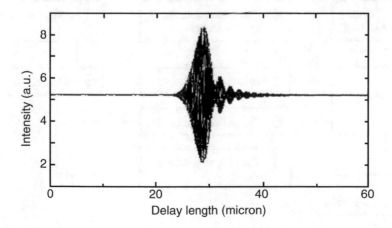

Fig. 3.1 Sample temporal interferogram

stable result rather than temporal interferometric. Two types of spectral interferometric are general case that called unbalance and special case that called balance. In
special case it is possible to measure the dispersion directly, so the achieved result
is more reliable than all above.

3.3   Apparatus

In the following sections, the main features and abilities of devices are inspected.
Tuneable laser, coupler, digital oscilloscope, external modulator, single-mode fibre,
polarization controller, coupler, laser diode, power supplier and function generator
would be studied in advance. Most of the information in the following sections are
taken from factory's data sheet and might be unique for specific product. As a result
due to available tolerance on device for different products, the achieved results
would have minor differences on measured values rather than desirable values.

3.3.1   Tuneable Laser Source (TLS)

The block diagram structure of tuneable laser model AQ8201-13 for presenting the
operation and controlling process is illustrated in Fig. 3.2 [10].

As it shows in above figure, in block number one, the signal control from the
frame is managed by the controller block. Block number 2 provides the ability of
changing wavelength by using pulse motor. In block 3, based on the position of the
switch, the laser would be on or off. In block 4, sub-multi-assembly (SMA) connec-
tor is used as an input port for modulation signal.

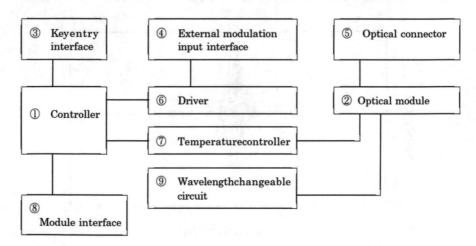

Fig. 3.2 The connectivity and functionality of each part in tuneable laser module AQ8201-13

Table 3.1 AQ8201-13 main features [10]

Wavelength range	1460–1580 nm
Wavelength resolution	10 pm
Optical output level in the range 1520–1580	+10 dBm
Operating temperature	23 ± 5
RIN	−145 dB/Hz
Optical connector	FC/PC

Block 5 is the final stage interface that the laser beam is radiated out. Block 6 is sort of circuit for driving the optical module. Since the output laser intensity is the function of temperature, block 7 is responsible for controlling temperature. Thus, the intensity amplitude of laser at the normal condition is approximately stable and constant. For manual controlling purpose and conducting the input data from the frame, the module interface in block 8 is employed.

The driver circuit for dealing with the pulse motor (related with changing wavelength in block 2) is the task of block 9. The considerable features of AQ8201-13 are gathered in the following (Table 3.1):

3.3.2 Polarization Controller (PC)

The main purpose of using *PC* for this experiment is related with involving external modulator. It is required to adjust the polarization state of laser in the fibre before it enters to the external modulator. The birefringence phenomenon helps for controlling polarization and can be applied by coiling fibre.

The induced retardation is corresponding with two factors, first the length of fibre and the second inverse of bending radius. The *FPC 560* is one of the popular polarization controllers and classified as "*BAT EAR*" controllers because it uses some holders at specific diameters and is adjustable for certain number of fibre turns in order to induce retardation.

Each paddle in Fig. 3.3 can make effective retardation corresponding to the number of fibre turns at specific diameter (if the paddle diameter is too small, the bending loss could happen). For *FPC560*, the first, the second and the third paddles make λ/4, λ/2 and λ/4 retardation, respectively. By changing the angle of first paddle, the input polarization state converts into linear polarization state, and the third state would convert it into the arbitrary output polarization state at a fixed wavelength. Since the polarization state is a function of wavelength, the high peak power can be achieved due to polarization rotation. Figure 3.4 depicts how lower wavelengths and higher number of loop retardation take the greater values compared with higher wavelengths [11].

Main specifications for *FPC560* are as follows:

- Paddle material: Black Delrin
- Loop diameter: 2.2″ (56 mm)
- Paddle rotation: ±117.5°
- Foot print (W × L): 1.0″ × 12.5″

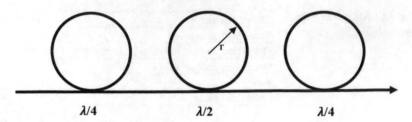

Fig. 3.3 *"BAT EAR"* polarization controller

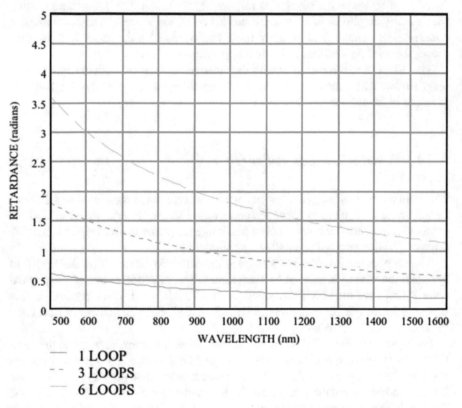

Fig. 3.4 Retardation vs. wavelength for 1, 3 and 6 fibre loops per paddle. The fibre clad diameter is 80 μm

3.3.3 Lithium Niobate (LiNbO₃) Electro-optic Modulator (EOM)

The light characteristics can handle many applications such data transmission, sensor, measuring, etc. One of the popular devices that can help to exploit these characteristics is optical modulator. Since light has many variable parameters to manipulate by optical modulator, there are many types of modulators such as phase modulator, intensity modulator, etc. [12].

Since debating about available types of modulator is beyond the scope of this book, just Mach-Zehnder which is employed for this experiment would be studied. Mach-Zehnder modulator is usually categorized as interferometric modulators and in terms of application mostly used for optical data transmission. The employed modulator for this experiment is called "F10", or "low drive voltage 10–12.5 Gb/s modulator small form factor chirp-free", that it represents the main features of this modulator.

The "small form factor" expression is usually used for those components that have the capability of enhancing port density, and also their packaging must support *small form factor* (SFF) technology. "Low drive voltage" shows the fact it requires very low voltage to operate at very high-speed transmission.

This modulator is chirp-free; in other words, it means after modulation, there is no change in frequency of optical signal induced by modulator with time. Figure 3.5 shows different modulation schemes using $LiNbO_3$ [13].

Electro-optic coefficient provides the ratio of electrical to optical conversion. There are many types of electro-optic coefficients, and r_{33} provides the highest among

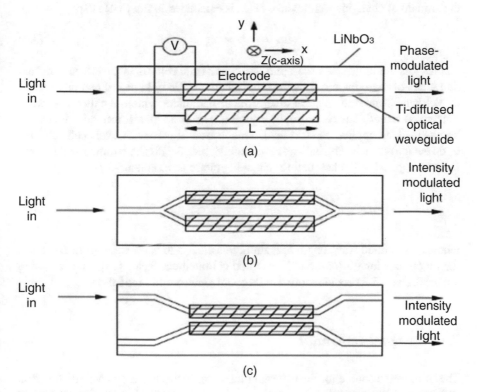

Fig. 3.5 Pattern of z-transect $LiNbO_3$ waveguide. (**a**) Phase modulator, (**b**) intensity modulator and (**c**) directional coupler

Table 3.2 General operating conditions

Operating case temperature (Top)	−5 to +75 °c
Storage temperature	−40 to +85 °c
Maximum top variation rate	1 °c/min
RF input power (electrical)	25 dBm (AC coupled)
Optical input power	20 dBm (continuous wave)

Table 3.3 Specific operating conditions

Operating wavelength	1525–1605 nm
Optical insertion loss (2)	5 dB
RF V_π voltage	3.8–5 V at 1 kHz
Bias V_π voltage	5.5–6 V at 1 kHz
Optical return loss	45 dB
DC optical extinction ratio	20–24 dB
Chirp, alpha parameter	−0.1–0.1

of them [14]. In the condition that the electrodes were placed on top of the waveguide arm in Z-transect wafer and X-transect wafer was placed on both sides of the branches, the amount of changing in refractive index is calculated in Eq. (3.4) [13]:

$$\Delta n_e = 0.5 \cdot n_e \cdot r_{33} \cdot E_z \qquad (3.4)$$

where r_{33} is the highest electro-optic coefficient (pm/Volt), n_e is the effective refractive index of material and E_z is the propagated electric field in Z-direction in terms of Volt/cm. Consecutively, the other critical parameter which is called half-wave voltage is studied. The electro-optic crystal is a medium that its refractive index can be changed by applied electric field. This crystal is part of Pockels cell (electro-optic device) device. The half-wave voltage V_π (Eq. 3.5) is the required voltage that must be applied to Pockels cell for changing phase as π radian.

$$V_\pi = \frac{\lambda \cdot g}{L_m \cdot n_e^3 \cdot r_{33} \cdot \Gamma} \qquad (3.5)$$

where λ is defined wavelength operation, g is electrode gap length (μm), L_m shows the modulator length (cm) and Γ is called confinement factor [13]. The following Tables 3.2 and 3.3 provide more details about electro-optic modulator.

3.3.4   Signal Generator

The microwave source for the external modulator is produced by PM5191 function generator. It has capability to produce different waves like sine and square form. The frequency domain also covers the range between 0.1 MHz and 2 MHz. Accuracy and stability are the essential factors for produced sine wave that fulfils with this synthesizer/function generator. Another important feature of PM5191 is remote programmability using GPIB port. All front panel functions can be programmed remotely, and all settings and status data can be called and recalled through the remote controller. Some major features about the function generator are as follows:

Sine wave	0.1 MHz–2.147 MHz
Square wave	0.1 MHz–2.147 MHz
Phase noise	<−80 dBc/Hz
Signal to noise ratio (SNR)	≥55 dBc
Long-term drift	<0.3 ppm within 7 h
Max resolution	0.1 MHz
Voltage peak-to-peak open circuit (for both sine wave and square wave)	0–30 V
Rise and fall time for square wave	<35 ns
Duty cycle	50%

3.3.5 Optical-to-Electrical Converter (O/E)

In the field of optic, for the application of measuring signal and further analysis, sometimes it is required to convert optical form into microwave form. For the purpose of this work, Agilent 11982A is utilized by means of PIN photodetector to convert optic signal to electric form. For optical application, the PIN photodetector can support bandwidths of the order of tens of gigahertz [14]. After detection process, the produced electrical signal needs to be amplified by very low-noise amplifier due to high attenuation during conversion. The amplified electrical signal is suitable to use with measuring instrument such as digital oscilloscope or spectrum analyser. In frequency domain, the optical characteristics of laser beam such as intensity modulation, distortion and laser intensity noise can be displayed by spectrum analyser, and the effect of laser modulation is measurable. Another important capability of the 11982A is changing display oscilloscope unit to watt unit in order to measure optical power. For this purpose it is enough to enter the reciprocal of the responsivity in oscilloscope's probe attenuation field. Considerable features of the O/E 11982A are as follows [15]:

Wavelength range	1200–1600 nm
Bandwidth optical	DC to 15 GHz
Bandwidth electrical	DC to 11 GHz
Input return loss	23 dB
Conversion gain (accuracy of provided value)	±20%
Maximum safe optical input power (average)	10 mW
Temperature range	0 °C–+55 °C

3.3.6   Optical Multimetre

To monitor the measured optical power and wavelength metre simultaneously, the OMM-6810B is applied for this experiment. This model is used for general laboratory purpose and can be used for laser diode and any other laser source. It also supports GPIB/IEEE 488.2 interface which enables us to measure two important parameters, optical power and wavelength, automatically. More considerable experiment features are as follows [16]:

Range linear power	0.000–999.99 W
Range log power	−99.999–99.999 dBm/dB
Range wavelength	190–30,000 nm
GPIB	Meets ANSI/IEEE Standard 488.1–1987
GPIB	Meets ANSI/IEEE Standard 488.2–1987
Operating temperature	10–40 °C

3.3.7   Digital Signal Oscilloscope

The "DSO1022a", from Agilent.co, has been employed to analyse the converted optical signal to microwave signal in time domain. This device supports two input RF channels. Automatic measurement function provides the feasibility of 22 automatic measurements. As shown in Fig. 3.6a, b, most of the voltage components and time components can be measured by DSO1022a [17].

Noticeable features of the DOSC can be summarized as follows:

1. Displaying automatic measurements for three parameters at the same time.
2. Automatically measuring ten parameters of voltage.

   (a) Maximum voltage
   (b) Minimum voltage
   (c) Peak-to-peak voltage
   (d) Top voltage
   (e) Base voltage
   (f) Amplitude voltage
   (g) Average voltage
   (h) Root mean square voltage
   (i) Over shoot
   (j) Pre-shoot

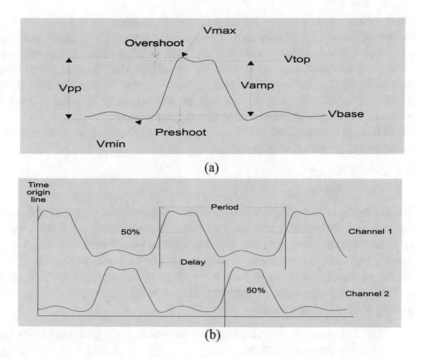

Fig. 3.6 (a) Measurable voltage components and (b) measurable time components between two channels in DSO1022a

3. Automatically measuring 12 parameters of time.

    (a) Period

    (b) Frequency

    (c) Rise time

    (d) Fall time

    (e) + pulse width

    (f) Pulse width

    (g) + duty cycle

    (h) Duty cycle

    (i) Delay A–B, rising edges

    (j) Delay A–B, falling edges

    (k) Phase A–B, rising edges

    (l) To calculate the delay between source one and source two, the conversion formula (3.6) is applied to the phase A–B, rising edges:

$$\text{Phase} = \frac{\text{Delay}}{\text{Source 1 Period}} \times 360^{\circ} \qquad (3.6)$$

    (m) Phase A–B, falling edges

4. Using cursor measurements for the purpose of measuring in the selectable area for both time and amplitude axes. This cursor can be set in manual, automatic and track mode. Manual mode lets the user to adjust the cursors manually. Automatic mode adjusts the cursors by the digital oscilloscope for the most recent displayed voltage or time measurements automatically. Track mode lets the user to adjust one or two cross hair cursors manually in order to track the points of wave form in both horizontal and vertical axes.

5. DSO1022a provides the ability of saving oscilloscope screen in terms of graphical or sampling data at two locations:

   (a) Internal storage
   (b) External storage

6. Trigger mode is activated when it is supposed the data is captured and stored. This mode includes some subfunctions. When the subfunctions are chosen properly, the unstable or blank screen is converted into preselected wave form. In the first trigger point, the oscilloscope captures enough data to draw the waveform in the left of the trigger point, and when the next trigger condition takes place, the process of collecting data would be continued to the right of the trigger point. Subfunctions of trigger mode:

   (a) Edge: to take place when the trigger input passes through a specified voltage level with the specific slope
   (b) Pulse: to search certain width pulses between other pulses
   (c) Video: to trigger on lines for standard video waveforms
   (d) Pattern: to trigger on patterns from all input channels
   (e) Alternate: to trigger on non-synchronized signals

7. Main features and environmental conditions of DSO1022a are collected as follows:

   (a) Bandwidth 200 MHz
   (b) Time base accuracy: ±50 *part per million* (ppm) from 0 °C to 30 °C
   (c) Trigger sensitivity for both channels: ≥5 mV/div – 1 div from DC to 10 MHz and 1.5 div from 10 MHz to full bandwidth
   (d) Peak detection: 4 ns
   (e) Coupling: AC, DC, ground
   (f) DC vertical gain accuracy: 2–5 mV/div – ±4.0% full scale

3.3.8   SMF 28

Single-mode optical fibre "SMF-28" has set the standard for value and performance for telephone, cable television, submarine and utility network applications. It is widely used in the transmission of voice, data and/or video services. It is manufactured to the most demanding specifications in industry. The following specifications can be considered as main features of SMF-28:

- Versatility in 1310–1550 nm applications
- Cable cut-off wavelength < 1260 nm
- Core diameter 8.2 μm
- Numerical aperture: 0.14
- Effective group index of refraction: 1.4682 at 1550 nm
- Attenuation: <0.22 at 1550 nm
- Zero-dispersion wavelength: 1313 nm

3.4 Summary

At the first and second sections of this chapter, some methods of dispersion measurement have been studied as a start point; consecutive sections review basics and main features of laboratory's apparatus. In the third section, a set of optical devices such as external modulator and measuring devices such as digital oscilloscope and so on are reviewed in advance. Since one of the main outcomes of this book is auto-measuring and remote controlling the capable devices, it is essential to know their functions.

References

1. J.H. Wiesenfeld, J. Stone, Measurement of dispersion using short lengths of an optical fiber and picosecond pulses from semiconductor film lasers. J. Lightwave Technol. **LT-2**, 464 (1984)
2. P. Merrit, R.P. Tatam, D.A. Jackson, Interferometric chromatic dispersion measurements on short lengths of Monomode optical fiber. J. Lightwave Technol. **7**, 703–716 (1989)
3. Y.O. Noh, D.Y. Kim, S.K. Oh, U.C. Pack. Dispersion measurements of a short length optical fiber using Fourier transform spectroscopy, *ThB5, Cleo, Pacific Rim'99*, pp. 599–600, (1999). P. J. Harshman, T. K. Gustafson, P. Kelley, Title of paper, J. Chem. Phys. **3**, (to be published)
4. P. Hamel, Y. Jaouen, R. Gabet, Optical low-coherence reflectometry for complete chromatic dispersion characterization of few-mode fibers. Opt. Lett. **32**(9), 1029 (2007)
5. F. Hakimi, H. Hakimi, Measurement of optical fiber dispersion and dispersion slope using a pair of short optical pulses and Fourier transform property of dispersive medium. Opt. Eng. **40**(6) (2001)
6. C. Palavicini, Y. Jaouën, G. Debarge, E. Kerrinckx, Y. Quiquempois, M. Douay, C. Lepers, A.-F. Obaton, G. Melin, Phase-sensitive optical low-coherence reflectometry technique applied to the characterization of photonic crystal fiber properties. Opt. Lett. **30**, 361 (2005)
7. A. Wax, C. Yang, J.A. Izatt, Fourier-domain low-coherence interferometry for light-scattering spectroscopy. Opt. Lett. **28**, 1230–1232 (2003)
8. K. Takada, I. Yokohama, K. Chida, J. Noda, New measurement system for fault location in optical waveguide devices based on an interferometric technique. Appl. Opt. **26**, 1603–1605 (1987)
9. R.K. Hickernell, T. Kaumasa, M. Yamada, M. Shimizu, M. Horiguchi, Pump-induced dispersion of erbium-doped fiber measured by Fourier-transform spectroscopy. Opt. Lett. **18**(1), 19–21 (1993)

10. *AQ8201 Series Optical Test& Measurement System Instruction Manual* (Andoelectric Co. Ltd, Japan, 2000), p. 200
11. *Fiber Polarization Controller-FPC560 Data Sheet*
12. M. Jarrahi, T.H. Lee, D.A.B. Miller, Wideband, low driving voltage traveling-wave Mach–Zehnder modulator for RF photonics. IEEE Photonic Technol. Lett. **20**(7), 517–519 (2008)
13. A.E.N.A. Mohamed, M.A. Metawe'e, A.N.Z. Rashed, A.M. Bendary, Recent progress of LiNbO3 based electrooptic modulators with non return to zero (NRZ) coding in high speed photonic networks. Int. J. Inform. Comm. Technol. Res. **1**(2), 55–63 (2011)
14. G.L. Li, S.A. Pappert, C.K. Sun, W.S.C. Chang, and P.K.L. Yu, *Wide Bandwidth Travelling Wave InGaAsP/InP Electro Absorption Modulator for millimeter Wave Applications*, in *IEEE MTT-S Int. Microwave Symp. Dig.*, 2001, pp. 61–64
15. TIA/EIA FOTP-175, *Chromatic dispersion measurement of single mode optical fibers by the differential phase shift method* (Telecommunications Industry Association, Washington, 1992)
16. *User's guide, optical power and wavelength meter OMM-6810B*
17. *Agilent 1000 Series Oscilloscopes, programmer's guide*

Chapter 4
Design and Development of Algorithm for Auto-Measurement Voltage and Temporal Parameters of Microwave Signal

4.1 Overview

This chapter presents the main work of this book. An optical device must be programmed to allow auto-measuring and provide high accuracy and reliability of the acquired results. In the following sections algorithms, charts and practical setups for characterizing a low drive voltage modulator and dispersion measurement systems are studied.

4.2 Experimental Setups for Characterizing a Tuneable Laser

In the setup shown in Fig. 4.1, two programmable optical devices, a tuneable laser and an optical multimeter, are connected to a personal computer (PC) or another programmable device that supports LabVIEW software and a general purpose interface bus (GPIB).

Both optical devices and the PC are equipped with GPIB ports. The PC as a controller device is responsible for analyzing the data received from both devices and sends necessary commands for controlling purposes.

4.3 Flowchart to Characterize the Tuneable Laser

For characterizing the tuneable laser, two mechanisms must be executed simultaneously. First, LabVIEW checks, step by step, every necessary data item gathered and then sends the command for initializing the tuneable laser. VISA function has an

© The Author(s), under exclusive license to Springer Nature Switzerland AG 2019   25
I. S. Amiri, M. Ghasemi, *Design and Development of Optical Dispersion Characterization Systems*, SpringerBriefs in Electrical and Computer Engineering, https://doi.org/10.1007/978-3-030-10585-3_4

Fig. 4.1 Experimental
setup for characterizing
tuneable laser

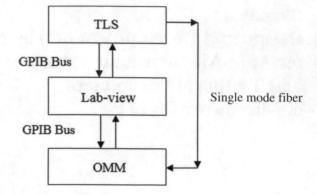

important role for connecting LabVIEW to external hardware. For many tasks, such as identifying weather, the device is connected to or disconnected from external ports [GPIB, universal serial bus (USB), etc.]; exporting and importing data from or into the device are tasks of VISA functions.

VISA open function is considered as the property of the VISA family function to open a session to the device specified by the VISA resource name; it returns a session identifier that can be used to call any other operations of that device. However, it has tasks such as VISA open time out, VISA resource name and VISA access mode; for the purpose of our experiment, however, using this property is fundamental.

The other important function is the VISA write function. This function writes the data from the write buffer to the device or interface specified by VISA resource name and any programmable device, and somehow needs to function to operate in the remote condition. Controlling TLS required defining some parameters such as start point, ending point, and steps to determine at which range of wavelength the Tunable Laser Source (TLS) must operate.

After initialization, a command for changing wavelength with specified time delay is issued. Because of the chosen step this routine would be continued until the end number for the wavelength (Fig. 4.2a). An optical multimeter tracks the received laser beam and measures its power and wavelength in a synchronous manner. To synchronize both the TLS and optical multimeter (OMM) there are different methods such as hand shaking, but here these two devices would be synched by the time delay. In other words, each device works independently but within a specific time delay.

The lower complexity of programming is considered as one advantage and lesser accuracy another advantage of the time synchronization method because for any device foulness or interruption cannot be detected by another device or Lab-view. Figure 4.2b illustrates the progress of programming for the OMM in a very short and simple scheme.

The GPIB write function has been employed by Lab-view to write data to the GPIB device that is identified by the address string subfunction. This function is also contains other resources such as "address string," which contains the address of

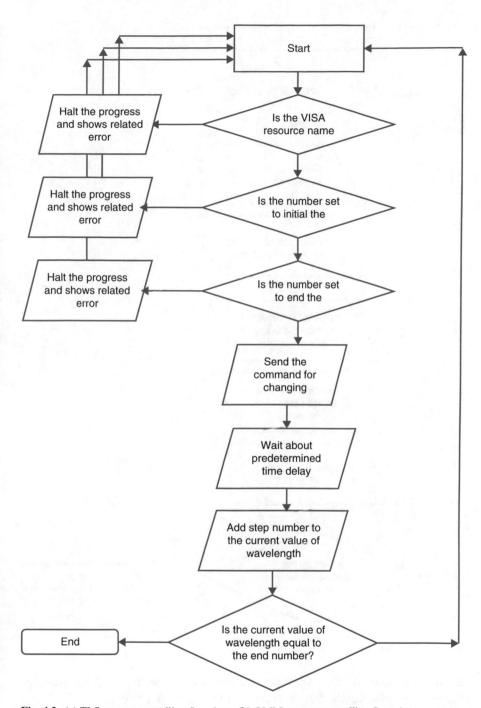

Fig. 4.2 (**a**) TLS remote controlling flowchart. (**b**) OMM remote controlling flow chart

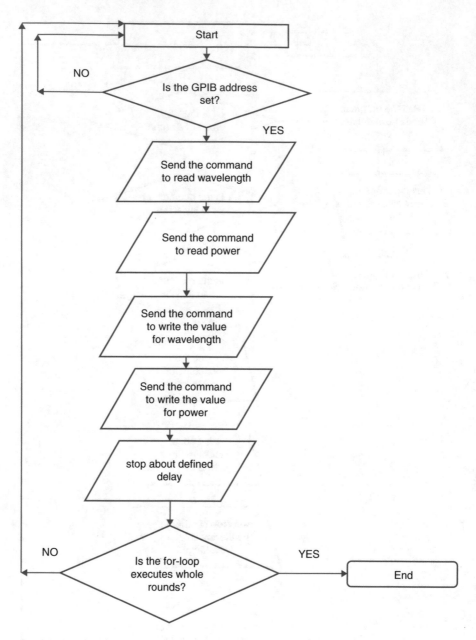

Fig. 4.2 (continued)

the GPIB device with which the function communicates, or "timeout," that specifies the time, in milliseconds, that the function waits before timing out.

Depending on transferred command, the response from the device is the output of the GPIB write function and is called "status:" it is a Boolean array in which each

Table 4.1 GPIB write function, numeric value and symbolic status

Numeric value	Status bit	Symbolic status	Description
1	0	DCAS	Device Clear state
2	1	DTAS	Device Trigger State
4	2	LACS	Listener Active
8	3	TACS	Talker Active
16	4	ATN	Attention Asserted
32	5	CIC	Controller-in-Charge
64	6	REM	Remote State
128	7	LOK	Lockout State
256	8	CMPL	Operation Completed
4096	12	SRQI	SRQ Detected while CIC
8192	13	END	EOI or EOS Detected
16,384	14	TIMO	Timeout
−32,768	15	ERR	Error Detected

bit describes a state of the GPIB controller. There are different conditions for status. The numeric value and symbolic status of each bit in status are collected in the following table with a description of each bit (Table 4.1).

The result from GPIB write function status needs to be converted from string to numeric for mathematical manipulation and labelled on the basis of the measured unit.

4.4   Experimental Setup for Characterizing the Electro-optic Modulator (EOM)

To characterize the external modulator, some active and passive devices are connected to provide the ideal situation for EOM performance. The TLS, as was mentioned in the previous section, is controlled by Lab-view and produces a laser beam in a certain wavelength at a specific power. To compensate for the depolarization loss caused by the imperfection of the fibre structure, a polarization controller must be used. Then again, EOM is very sensitive to the polarization state of the incoming optical beam (Fig. 4.3).

The type of EOM for the current experiment is classified as a Mach-Zehnder interferometer. It uses the splitting and recombination of the optical beam method for manipulating the characteristics of light such as phase or frequency. The two arms for the purpose of beam splitting might have provided different pathlengths [1], but if we assume it has the same optical pathlength the distribution of total power on the output may not change very much after recombination.

The other significant matter is the very high intrinsic loss of EOM. The inherent loss and depolarization loss together make a dramatic reduction of the received optical power. As mentioned earlier, by using a PC it is possible to compensate the

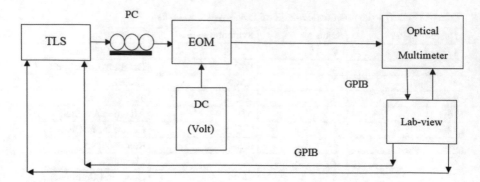

Fig. 4.3 Block diagram for measuring radio-frequency (RF) parameters received from the electro-optic modulator (EOM)

depolarization loss, but the EOM inherent loss is fixed and cannot be changed. To put the EOM in ideal operating point, the programmable DC power supply is employed. These devices must be treated manually, and for this reason the whole setup is not fully utilized in the order of auto-measurement.

Adjusting the PC paddle as a passive device has a negative effect on the performance of auto-measurement because it is necessary to ensure that the highest optical power is received and measured by the optical multimeter; then, measurements can begin. It is essential to characterize the EOM before and after applying in any setups because it might give a different response regarding the applied DC voltage and RF waves. Before the tuneable laser and optical multimeter start their algorithm under the vision of Lab-view, it is essential to ensure the absence of RF voltage over EOM pins because it shifts the EOM DC bias operating point and the achieved result is not accurate. When it is ensured that the highest power is being received by the OMM, the required conditions for starting characterization of EOM are confirmed, and the TLS and OMM algorithm can be run by Lab-view software. The DC power supplier is manually programmable and produces voltage at equal intervals of time like a counter.

The DC power supply cannot be controlled remotely and for the purposes of characterization the OMM must operate with DC power in the same time; thus, the only way to synch these two devices is by time synchronization. However, this solution decreases the efficiency of measurement in terms of time, although it is usable. First, a delay time must be defined for one device so that the other device can finish its task in time; then, the delayed device should proceed to the next step while the other device is delayed. The following Fig. 4.4 depicts the time synchronization of two devices. One considerably negative aspect of this method is the probable desynchronization of two devices as the result of any failure or bad starting point occurring. The time delay is fixed, and no physical feedback is provided to inform each of the devices. Here, it would be better for the DC power supply as a resource to start to work first, and then the OMM.

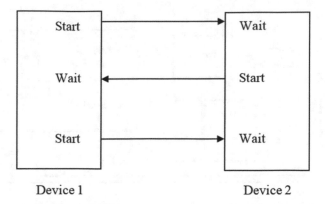

Fig. 4.4   Time synchronization of two devices

4.5   Experimental Setup for Measuring Critical Parameters of Received Optical Power

After characterization of EOM in terms of determining the DC bias operating point, it is then possible to apply the RF wave to the EOM to modulate the data on an optical carrier. As mentioned in the previous section, the employed function generator could not be controlled remotely, and voltage, frequency, and mode of pulse shape need to be entered manually.

The optical wave after EOM is converted to the form of a microwave wave by the optical to electrical converter in such a way that the digital oscilloscope is able to measure the microwave components in terms of voltage and frequency of modulated data and also the difference phase between original microwave and the received microwave after conversion. These parameters can be measured automatically by Lab-view because this device has the ability of remote controlling. The block diagram for RF parameter measurement is illustrated in Fig. 4.5. Because the routine of programming for making measurements has sequential progress, some flowcharts in the next figures present the measurement algorithm in a few sequences.

Figure 4.6 illustrates the flowchart in which Lab-view collects the external information before starting the measuring algorithm. There is no need to use a long length of fibre to characterize EOM, but this field is required to be filled for the next applications: the amount value for applied DC bias voltage and peak-to-peak AC voltage.

The frequency must be provided manually because there is no direct connection between Lab-view and unsupported remote controlling devices. The scenario in sequence two starts by sending a command to the digital oscilloscope to put it in the mode of acquiring data. In acquisition mode, the digital oscilloscope stops all current progress and responds to the request for capturing data from a front panel key or remote controller device.

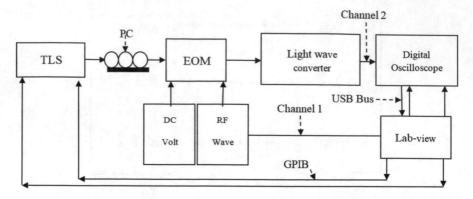

Fig. 4.5 Block diagram for measuring RF parameters of received optical power

It is essential to check that the delay time for acquisition mode does not exceed a certain duration, or else the progress must be halted and the exceeded time error issued to inform the user to increase the delay for longer measurement. The other requirement for controlling purposes is continuous inquiry to check the acknowledgment query from the digital oscilloscope to determine whether all pending device operations are completed. Any failure to respond from pending devices leads issuing an error and halting the progress of communication between two devices. All these monitoring actions are the properties of VISA function and are an effective way to guarantee the security of connection. A list of measurement commands is provided in Table 4.2. (These commands for the current experiment are used by Lab-view and taken from the data sheet of DSO1022a.)

Each RF channel has its own voltage or time properties, and they must be treated separately, but for measuring phase or delay between two channels the two channels must be evaluated concurrently. Hence, dependent on type of parameter either channels or both channels need to be associated; the flowchart in Figs. 4.7, 4.8, and 4.9 shows how we can achieve this purpose. The received results from channels regarding type of command are printed out in other formats such as Microsoft Excel or Word format or saved in a new variable for further analysis.

Because the measuring phase is one of the main objectives to observe dispersion, further mathematical analysis has to be taken over received measured phases. In sequence three of the measuring algorithm, because of the availability of a variety of mathematical functions, first the MATLAB script is a sort of interface software called MATLAB software. The MATLAB script invokes the MATLAB software script server to execute scripts written in the MATLAB language syntax: then, all the captured phases are indexed and employed by MATLAB software for statistical and mathematical operations, plotting results, etc. In sequence three the task of the digital oscilloscope has been finished and a new session would be started (Fig. 4.10).

A sequential algorithm is used because each sequence must be finished, and the results should be sent to the next sequence or, in other words, each sequence is dependent on the status of the previous sequence, except about sequence three, and

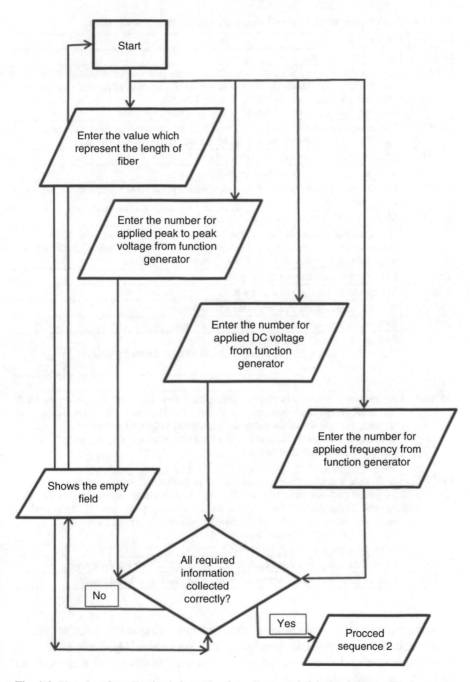

Fig. 4.6 Flowchart for collecting information from disconnected device (sequence1)

Table 4.2 Employed commands for second sequence of measurement algorithm

Command	Description
RIS?	Returns the rise time
FALL?	The fall time is determined by measuring the time at the upper threshold of the falling edge, then measuring the time at the lower threshold of the falling edge, and calculating the fall time with the following formula: fall time = time at lower threshold − time at upper threshold
FREQ?	Returns the measured frequency value in hertz of the first complete cycle on the screen using the mid-threshold levels of the waveform (in NR3 format)
NDUT?	Measures negative duty cycle in percent (%)
PDUT?	Measures positive duty cycle in percent (%)
NWID?	Returns the measured width of the first negative pulse using the mid-threshold levels of the waveform
PWID?	Measured width of the first positive pulse
VAMP?	Query returns the calculated difference between the top and base voltage. To determine the amplitude, the instrument measures Vtop and Vbase, then calculates the amplitude as follows: vertical amplitude = Vtop − Vbase
VAV?	Returns the calculated average voltage
VBA?	The measured voltage value at the base
VMAX?	Returns the measured absolute maximum voltage
VMIN?	Returns the measured absolute minimum voltage
VPP?	Returns the peak-to-peak voltage The peak-to-peak value (Vpp) is calculated with the following formula: Vpp = Vmax − Vmin
VRMS?	Returns the DC RMS voltage
NPHA?	A phase measurement is a combination of the period and delay measurements. First, the period is measured on source1. Then the delay is measured between source1 and source2. The edges used for delay are the source1 falling edge used for the period measurement closest to the screen left edge and the falling edge on source 2 The phase is calculated as follows: phase = (delay/period of input 1) × 360
PPHA?	A phase measurement is a combination of the period and delay measurements. First, the period is measured on source1. Then the delay is measured between source1 and source2. The edges used for delay are the source1 rising edge used for the period measurement closest to the screen left edge and the rising edge on source 2 The phase is calculated as follows: phase = (delay/period of input 1) × 360
NDEL?	Returns the delay between negative-going edges on source1 and source2
PDEL?	Returns the delay between positive-going edges on source1 and source2

it is not concerned with the accuracy of received data, or any external device failure. In this section, the setup and controlling and processing algorithm have been discussed in detail. The next chapter uses the same algorithm for measurement but with a different setup (Fig. 4.11).

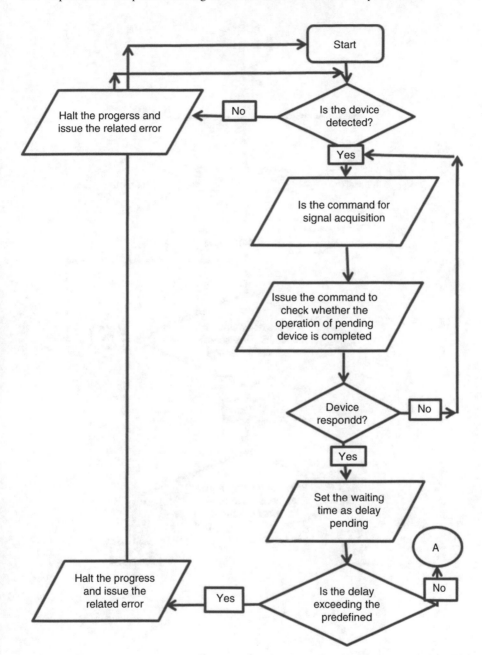

Fig. 4.7 Importing data from programmable digital oscilloscope (sequence 2)

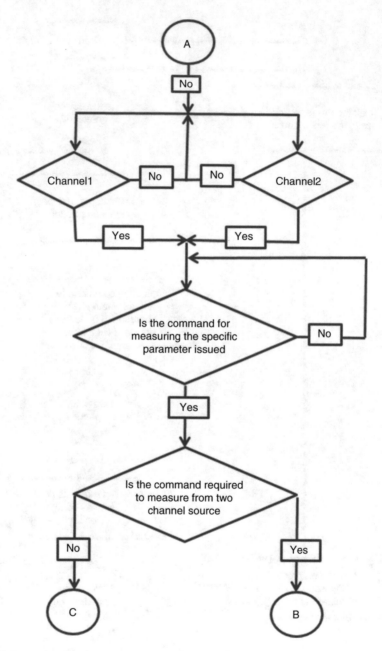

Fig. 4.8 Importing data from programmable digital oscilloscope (continuation of sequence 2)

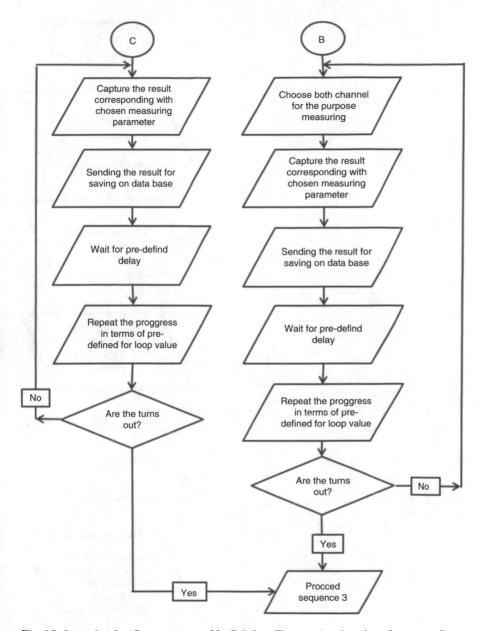

Fig. 4.9 Importing data from programmable digital oscilloscope (continuation of sequence 2)

4.6   Experimental Setup for Dispersion Measurement of Long-Length Fibre

The foregoing setup (Fig. 4.10) measures phase and other parameters for the received optical power in any length of optical fibre. There is not much difference between this setup and the setup in the previous section except the length of the fiber. All

Fig. 4.10 Indexing and
analyzing of captured data
(sequence 3)

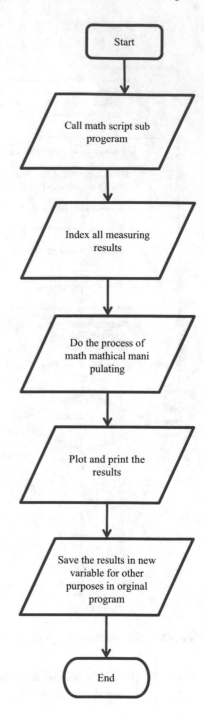

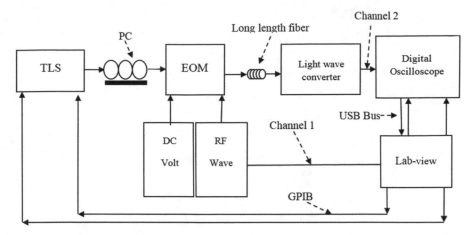

Fig. 4.11 Dispersion measurement setup for long-length fibre

examinations in the previous section are about the same as in the earlier setup. Some key points must be taken into account before and during the measurement.

1. At a different set of wavelengths the TLS produces the laser at different power. As a result for a greater length of fibre (more than some hundreds of kilometers), the boundary of the employed wavelength will be limited to a specific range (usually between 1540 nm and 1580 nm).
2. Before starting measuring, it must be ascertained whether the operating point of EOM and the applied peak-to-peak voltage are in the correct range.
3. The waveform on the digital oscilloscope display must be as similar as possible to the original microwave form, or the measured parameters related to dispersion cannot be trusted.

4.7 Summary

This chapter provides programming fundamentals and practical implemented set-ups, also explaining them in detail. Describing algorithms in terms of a flowchart merely shows the summary of a program for more clarification and how the program is developed. Real-time processing for very accurate results is essential, but because the amount of data for higher processing needs a very low delay response that is not provided by the hardware, thus those algorithms are developed based on a medium level of accuracy.

References

1. M.T. Jaekel, S. Reynaud, Quantum limits in interferometric measurements. Europhys. Lett. **13**, 301 (1990)

Chapter 5
Device Characterizations and Chromatic Dispersion Measurement in Optical Fibres

5.1 Overview

This chapter presents the measurement results obtained from the experiments that have been discussed in the previous chapter. The critical parameters such as wavelength, frequency and DC biasing voltage give different results based on assigned values. External modulator response also limits the dynamic range of these parameters. Since the effect of each parameter on dispersion can be studied separately, in each section of this chapter, each variable parameter will be measured, while the other parameter was made constant. In addition, analysis of multiple critical measurements will be presented to relate the experiment result with the theory.

5.2 Characterization of Tuneable Laser

Characterization concept is significantly essential for the resources due to their variation or shifting of output in time. By referring to Sect. 4.2, the setup for the purpose of characterization is presented. The TLS have to produce continuous wave in the way that received power from TLS at constant time.

Table 5.1 shows the results from measured power at random time and random wavelength. The below results show the TLS output optical level stability for over than 60 min would be changed but not considerably. Few hundredth changing of received optical power is tolerable for other parts of experiments.

© The Author(s), under exclusive license to Springer Nature Switzerland AG 2019   41
I. S. Amiri, M. Ghasemi, *Design and Development of Optical Dispersion Characterization Systems*, SpringerBriefs in Electrical and Computer Engineering, https://doi.org/10.1007/978-3-030-10585-3_5

Table 5.1 Characterization tuneable laser

Time (min)	Wavelength (nm)	Applied power (dBm)	Measured power (dBm)
10	1520	10.02	10.02
30	1540	10.02	10.02
60	1560	10.02	10.01
80	1540	10.02	10.01

5.3   Statistical Analysis to Find the Best Voltage Biasing Value Point of EOM

External modulator's requirement condition to use in the experiment is work on correct operating point. In the right operating point, the external modulator works ideally, and the data from RF port modulate on optical beam, and there is not much difference between original waveform and optical waveform except the amplitude. This section is referred to the previous chapter, Sect. 4.4, which is about characterization of external modulator. There are three types of EOM with different specifications. The characterization is executed several times, and the following figures in Fig. 5.2 are plotted. Ideally just one figure must be available as a reference figure, but due to variation of received optical power, it is necessary to make sure about the best point as operation point.

Table 5.2 depicts the result specifications at the laser power source around 7 dBm and wavelength 1544.5 nm with more details. Theoretically the Mach-Zehnder interferometer transmission transfer curve versus bias voltage is like Fig. 5.1. The two points in Fig. 5.1 are critical. The optical insertion loss occurred at these two points. V_π symbol represents the switching voltage and has two values at extremum points.

In approximate linear area (area between white colour arrow lines), the middle point which is between maximum and minimum point is the nominal operation point. For applied sinusoidal waveform to the EOM, the approx. peak-to-peak voltage should take the value between zero and the difference of maximum and minimum value's points until the sinusoidal optical waveform is the linear function of original RF sinusoidal waveform.

In positive area (the left-hand red solid line curve in Fig. 5.2), the sinusoidal waveform is in phase with original RF one, but in negative area (the right-hand red solid line curve in Fig. 5.2), 180° phase difference is available between them. The EOM response is linear for small fluctuations from the nominal operating bias point. The transmission light power $I(t)$ might be written as Eq. (5.1a) [1], if sinusoid waveform with frequency w is applied into RF port.

$$I(t) = \frac{I_0}{2}[1 + \cos\left(\pi \frac{V_b + E(w)V_p \cos(wt)}{V_\pi}\right)$$ (5.1a)

Table 5.2 Characterization of EOM results

(a)				
Model no.	Test no.	V_b (V)	V_π (V)	V_p (V)
Negative phase area				
EOM-7F6E1900	1	2–4	1.5–5.5	2
EOM-7F6E1900	2	2–4	1–5.5	2
EOM-7F6E1900	3	2–4	1.5–5.5	2
EOM-7F6E1900	4	2.5–4	1.25–5.5	1.5
Positive phase area				
EOM-7F6E1900	1	6.5–8.5	5.5–9	2
EOM-7F6E1900	2	6.5–8.5	5.5–9	2
EOM-7F6E1900	3	7–9	5.5–9.5	2
EOM-7F6E1900	4	7–9	5.5–9.5	2
Negative phase area				
EOM-7F6E1A00	1	3–4.5	2–6	1.5
EOM-7F6E1A00	2	3–4.5	1.5–6	1.5
EOM-7F6E1A00	3	2.5–4.5	1.5–5.5	2
EOM-7F6E1A00	4	2.5–4.5	1–6	2
Positive phase area				
EOM-7F6E1A00	1	7–8.5	6–10	1.5
EOM-7F6E1A00	2	7–8.5	6–9.5	1.5
EOM-7F6E1A00	3	7–8.5	6–9.5	1.5
EOM-7F6E1A00	4	7–8.5	6–9.5	1.5
(b)				
Positive phase area				
EOM-9F016502	1	5.5–7.5	4–8.5	2
EOM-9F016502	2	6–7.5	4–8.5	1.5
EOM-9F016502	3	5.5–7.5	4–8	2
EOM-9F016502	4	5.5–7.5	4–8.5	2

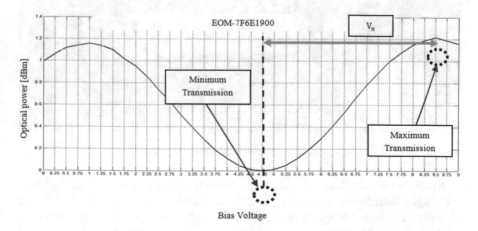

Fig. 5.1 EOM transmission transfer curves versus bias and related conceptions

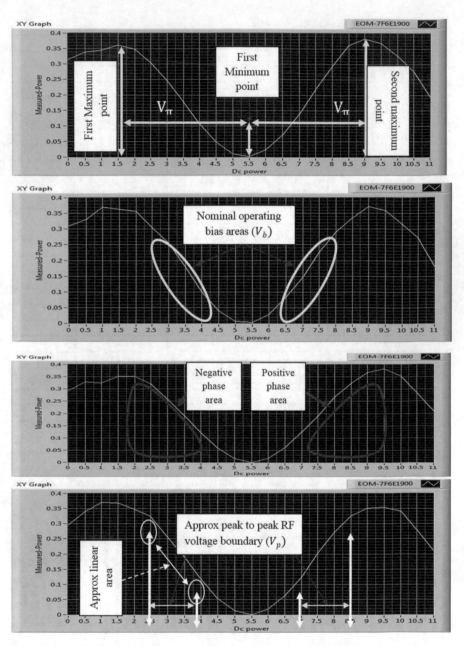

Fig. 5.2 Identification V_{p}, V_{b} and V_{π} over transmission transfer curves

In above formula, I_0 is the maximum transmitted light power, and the $E(w)$ represents the frequency dependency modulation efficiency. The two variables V_{p} and V_{π} are peak signal voltage to the EOM and modulator's characteristic switching voltage (from first minimum point to second maximum point, blue solid lines in

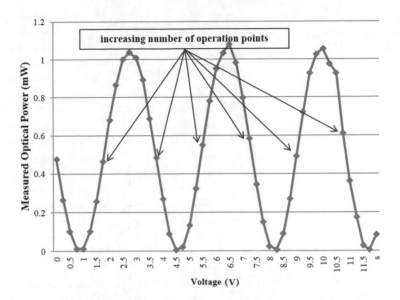

Fig. 5.3 Shifted transmission transfer curve

Fig. 5.2, and from first maximum point to the first minimum point, orange solid lines in Fig. 5.2, respectively).

The DC bias variable V_b (Eq. 5.1b) is an adjustable parameter and controls the operating points (Dennis Derickson et al. 1990). It is noticeable about shifting over transmission transfer curve when the out of defined range DC bias voltage is applied to the EOM. The curve might be shifts and the operation points are changed such as Fig. 5.3.

$$V_b = \frac{V_{max} + V_{min}}{2} \tag{5.1b}$$

The shifted EOM causes the linear operation area to become smaller; consequently the RF peak-to-peak voltage assigned very small values in linear operation bias area, and as a result, higher attenuation after long-distance fibre provides difficulties of detecting signal at the receiver.

To find the best operation point, the laser with the specifications in Table 5.3 is applied in the following figures that are captured that shows EOM response to the variation of DC bias voltage when the sinusoidal RF wave is passing through the RF port. It is clear the bias operation point is moved in transmission transfer curve [2] (Figs. 5.4, 5.5, 5.6, 5.7, 5.8, 5.9, 5.10, 5.11, 5.12, 5.13, 5.14, and 5.15).

The above figures clearly illustrate the response of EOM for different values of V_b. Since the boundary of linear operation bias is not just limited to a specific point, the highest similarity between the original waveform and measured waveform can be used as criterion of bias operation point. For example, for the first four figures, if

Table 5.3 New characterization to evaluate the variation of DC bias point

Wavelength	1540 nm
Power	10.02 dBm
EOM model No.	7F6E1900
Applied RF wave (volt)	1 V
Frequency	50 Hz
Applied DC power (V)	Measured optical power (mW)
0	0.90916
0.5	1.3
1	1.72
1.5	2.12
2	2.4
2.5	2.35
3	2.11
3.5	1.74
4	1.25
4.5	0.73641
5	0.34734
5.5	0.08968
6	0.00282
6.5	0.10841
7	0.41129
7.5	0.88028
8	1.4
8.5	1.89
9	2.26
9.5	2.43
10	2.38

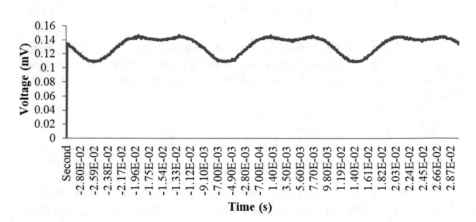

Fig. 5.4 Electrical form of received optical power at $V_b = 2.5$ V

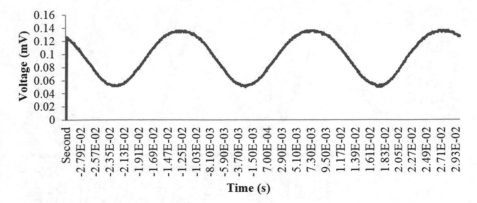

Fig. 5.5   Electrical form of received optical power at $V_b = 3$ V

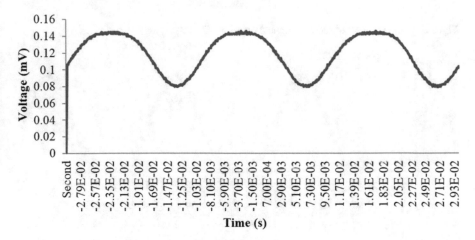

Fig. 5.6   Electrical form of received optical power at $V_b = 3.5$ V

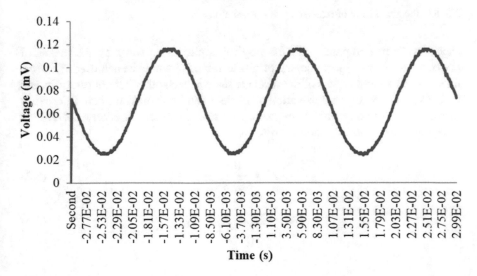

Fig. 5.7   Electrical form of received optical power at $V_b = 4$ V

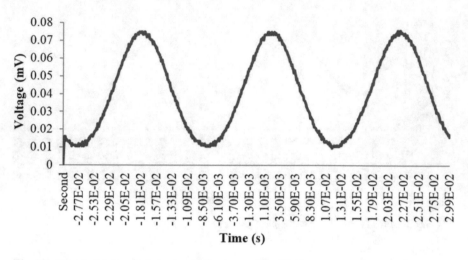

Fig. 5.8 Electrical form of received optical power at $V_b = 4.5$ V

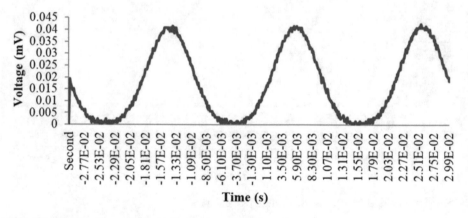

Fig. 5.9 Electrical form of received optical power at $V_b = 5$ V

each waveform compares with the original applied sine wave by the means of *Pearson's correlation* formula, the results in Table 5.3 would be achieved. This formula is used when two sets of data seem to scale to each other but not perfectly, and in this specific case, the correlation coefficient can be applied to them in order to recognize how two variables are correlated to each other. *Pearson's correlation* (Kreyszig 1999) formula can be written as

$$r^2 = \frac{SS_{xy}^2}{SS_{xx}SS_{yy}} \tag{5.2}$$

$$SS_{xy} = \sum (x - \bar{x})(y - \bar{y}) \tag{5.3}$$

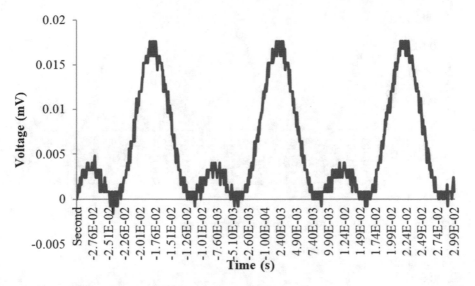

Fig. 5.10 Electrical form of received optical power at $V_b = 5.5$ V

$$SS_{xx} = \Sigma(x - \bar{x})^2 \qquad (5.4)$$

$$SS_{yy} = \Sigma(y - \bar{y})^2 \qquad (5.5)$$

where the x and y represent the voltage values of applied RF sine wave and measured RF wave, respectively. The average values of x and y also show as $\bar{x}$ and $\bar{y}$. Equations (5.4) and (5.5) indicate the sum of the squared deviation of x and y from their mean. Equation (5.3) indicates the multiplication of the deviation of x from its mean to the deviation of y from its mean.

For the correlation value near one, it means high correlation, while for the values near zero, it means poor correlation. The achieved results in Table 5.4 prove this fact; in the operation point of transmission transfer curve, the measured sine waveform has higher similarity to the applied sine waveform. In normal operation point, the EOM have the linear response, and how the V_b becomes closer to the linear point, the correlation coefficient becomes closer to the value number one. In addition, V_p is limited in the linear area, and at the higher amplitude in nonlinear area, the received waveform would be distorted and unreliable (Table 5.5).

The above table specifies the operation conditions, and the following figures show the results at different values of V_p in both linear and nonlinear areas of transmission transfer curve. It is clear that the upward changing voltage value causes the sine waveform step by step reshaped and distorted.

At the voltage amounts around 2.64 V, the sine waveform is acceptable and confirmed the fact that the EOM works in linear area, whereas, at the voltage around 4.16 V, the sine wave starts to deform in an irregular way, and roughly this is the start point in nonlinearity area of transmission transfer curve. At 9.36 V, it seems the frequency of received distorted sine wave became twice which is the extremum point on transfer curve (Figs. 5.16, 5.17, 5.18, 5.19, 5.20, 5.21, 5.22, 5.23, and 5.24).

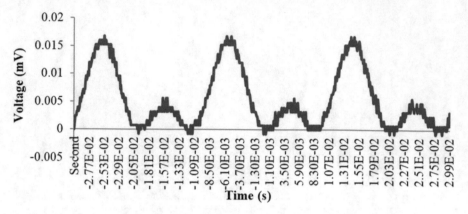

Fig. 5.11 Electrical form of received optical power at $V_b = 6$ V

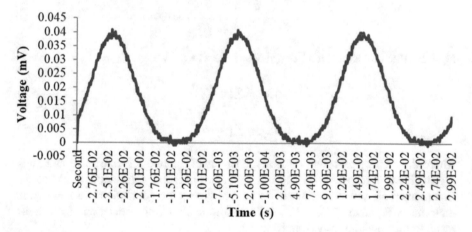

Fig. 5.12 Electrical form of received optical power at $V_b = 6.5$ V

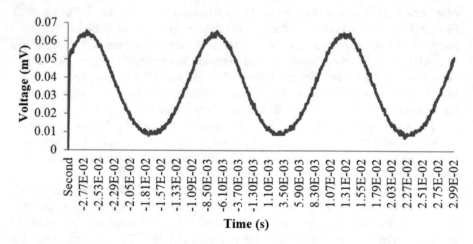

Fig. 5.13 (b) Electrical form of received optical power at $V_b = 7$ V

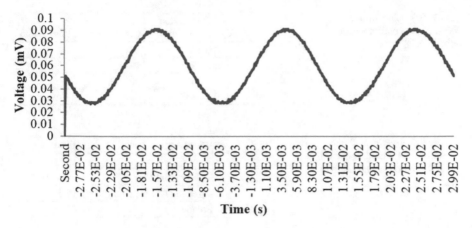

Fig. 5.14 Electrical form of received optical power at V_b = 7.5 V

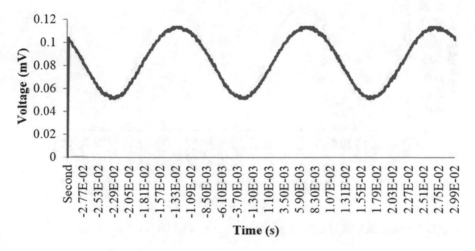

Fig. 5.15 Electrical form of received optical power at V_b = 8 V

5.4 Analysis of Measured Average Phase vs. Variable Frequency

This section studies the effect of low and high frequency on phase difference between electrical sine waveforms of received optical power from EOM and applied microwave electrical sine waveform to the EOM. To identify the relation between the phase and frequency of applied microwave signal, first the relation between output power and input power must studied. Equations (5.6)–(5.8) show how the ratio of EOM output power and input power is related with the applied RF signal [3–7].

Table 5.4 Calculating correlation coefficient for different V_b

V_b (V)	Correlation coefficient	Correlation coefficient (%)
2.5	0.1233	12
3	0.1301	13
3.5	0.2015	20
4	0.6835	68

Table 5.5 Operational condition for characterization of EOM based on variation of V_p

Wavelength	1540 nm
Power	10.02 dBm
EOM model No.	7F6E1900
Applied DC bias voltage (V)	7.5 V
Frequency	50 Hz

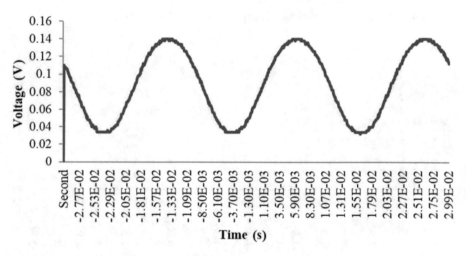

Fig. 5.16 Electrical form of received optical power at $V_p = 1.96$ V

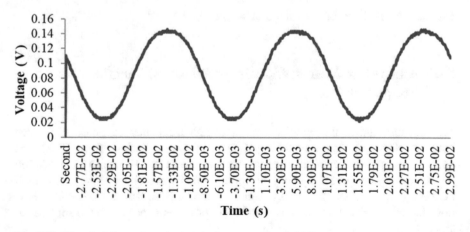

Fig. 5.17 Electrical form of received optical power at $V_p = 2.64$ V

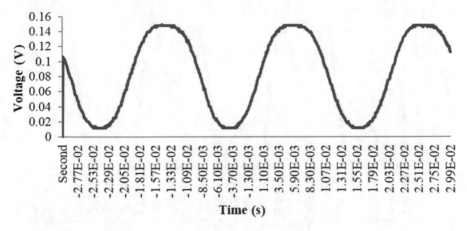

Fig. 5.18 Electrical form of received optical power at V_p = 3.64 V

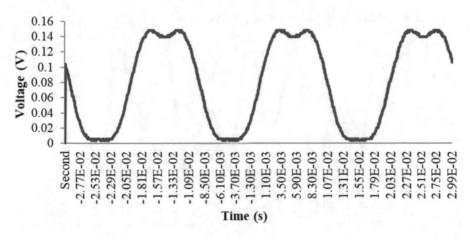

Fig. 5.19 Electrical form of received optical power at V_p = 4.16 V

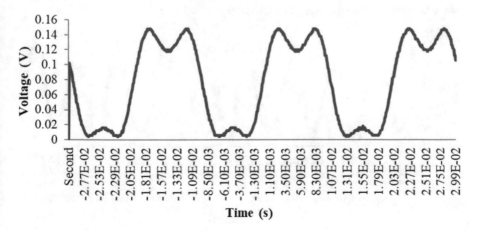

Fig. 5.20 Electrical form of received optical power at V_p = 5.28 V

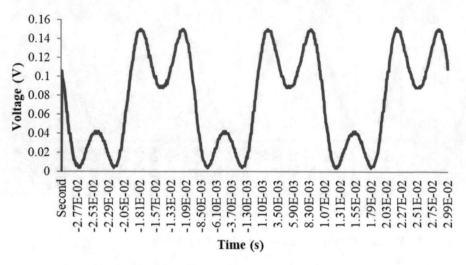

Fig. 5.21 Electrical form of received optical power at $V_p = 6.4$ V

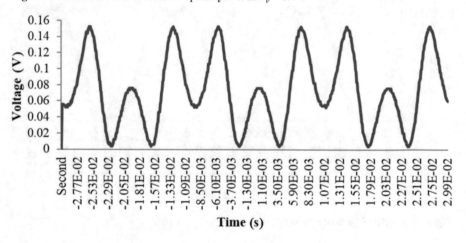

Fig. 5.22 Electrical form of received optical power at $V_p = 7.52$ V

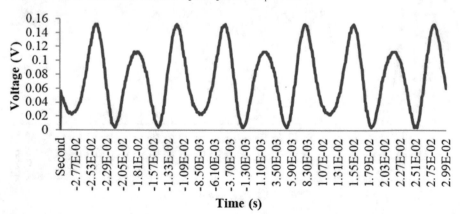

Fig. 5.23 Electrical form of received optical power at $V_p = 8.64$ V

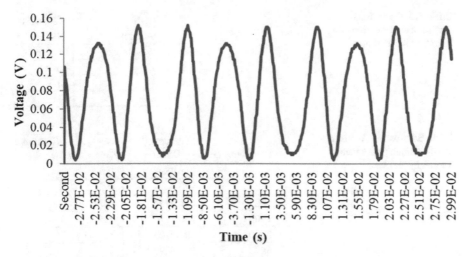

Fig. 5.24   Electrical form of received optical power at $V_\mathrm{p} = 9.36$ V

$$P_\mathrm{out} = 0.5\left\{\left(IE_\mathrm{A}I - IE_\mathrm{B}I\right)^2 + 2IE_\mathrm{B}I\cos^2\Delta\varphi\right\}$$

$$= 0.5P_\mathrm{in}\cdot\left(K_1 + K_2\cos^2\left(\frac{\pi V}{2V_\pi}\right)\right) \tag{5.6}$$

and

$$\Delta\varphi = \frac{\pi\cdot V}{2V_\pi} \tag{5.7}$$

In the absence of electrical signal, the two fields E_A and E_B that represent the input wave are divided into two arms equally. When $V = 0$ the combination of two fields would be constructive, but when $V = V_\pi$, the combination of two fields is destructive. By changing the applied voltage, the phase of waves in two arms would be changed, and at $V = 0$, the two waves are in the same phase, but at $V = V_\pi$, the phase difference is π, and the output light will become zero. The parameter V_π is also called driving voltage and dependent on wavelength and refractive index of LiNbO3 waveguide. Equation (5.8) shows how V_π is calculated (Rangaraj Madabushi et al. 2002). In Eq. (5.6), I is represented for optical intensity in each arm of EOM.

$$V_\pi = \frac{\lambda G}{2n_\mathrm{e}^3 r_{33}\Gamma L} = \frac{\lambda\cdot V}{4\cdot n(V)\cdot L} \tag{5.8}$$

where λ is the wavelength of applied laser beam, n_e is refractive index of LiNbO3 and r_{33} is the electro-optic coefficient ($3 \times 10^{-12}\,m/V$). The parameter V is the voltage applied,

Table 5.6 Operational condition specifications for the phase measurement

EOM	792000280
Serial no.	7F6e1900
Wavelength	1540 nm
V_b	8.5 V
V_p	1.1 V
Length	0 km

Table 5.7 Average measure of 100 sample phase versus frequency at 0 km

Frequency (Hz)	Frequency (dB Hz)	Average of 100 sample phase (°) = $\Delta\varphi$
50	16.9897	0.5
100	20	0.5
500	26.9897	0.7
1000	30	1
2000	33.0103	1.1
5000	36.9897	1.2
10,000	40	1.4
15,000	41.76091	1.7
20,000	43.0103	2.1
30,000	44.77121	2.2
60,000	47.78151	3
70,000	48.45098	3.6
80,000	49.0309	4
900,000	59.54243	4.6
100,000	50	5.7
200,000	53.0103	12.4
300,000	54.77121	21.6
500,000	56.9897	27.6
700,000	58.45098	30.5
800,000	59.0309	33.79
1,000,000	60	35.7
1,200,000	60.79181	39.39
1,400,000	61.46128	42.7
1,600,000	62.0412	45
1,800,000	62.55273	50.39
2,000,000	63.0103	56.6

and Γ is the overlap integral between optical and electrical (RF) fields, and G the free space gap between electrodes, and L is the electrode length. The following figures depict the measured phase, while the frequency and the length of fibre had been changed, and also Table 5.6 shows the specifications of laser, RF and DC biasing source.

Table 5.7 shows the measured average phase in terms of different frequencies which starts from very low frequency (50 Hz) to the high frequency (around 2 MHz) (Tables 5.8, 5.9 and 5.10).

Table 5.8 Average measure of 100 sample phase versus frequency at 2.2 km

Frequency (Hz)	Frequency (dB Hz)	First period $\Delta\varphi$ (°)	Second period $\Delta\varphi$ (°)	Third period $\Delta\varphi$ (°)
50	16.9897	0.5		
100	20	0.5		
500	26.9897	0.6		
1000	30	0.7		
2000	33.0103	5		
5000	36.9897	17.3		
10,000	40	37.2		
15,000	41.76091	56.29		
20,000	43.0103	78.3		
30,000	44.77121	116.2		
60,000	47.78151	235		
70,000	48.45098	275.3		
80,000	49.0309	318		
900,000	59.54243	354.9		
100,000	50	35.5	390.4	
200,000	53.0103	71.5	461.9	
300,000	54.77121	105.6	567.5	
500,000	56.9897	175.6	743.1	
700,000	58.45098	236.8	979.9	
800,000	59.0309	272.1	1252	
1,000,000	60	328.7	1580.7	
1,200,000	60.79181	28.2		1608.9
1,400,000	61.46128	90.5		1699.4
1,600,000	62.0412	147.1		1846.5
1,800,000	62.55273	210.8		2057.3
2,000,000	63.0103	264.9		2322.2

Table 5.9 Average measure of 100 sample phase versus frequency at 12.2 km

Frequency (Hz)	Frequency (dB Hz)	First period $\Delta\varphi$ (°)	Second period $\Delta\varphi$ (°)	Third period $\Delta\varphi$ (°)
50	16.98970004	4.3		
100	20	4.9		
500	26.98970004	10		
1000	30	20.5		
2000	33.01029996	37.9		
5000	36.98970004	99.5		
10,000	40	214.9		
15,000	41.76091259	318.6		
20,000	43.01029996	71.5	390.1	
30,000	44.77121255	284.4		674.5

Table 5.10 Average measure of 100 sample phase versus frequency at 37.2 km

		First period	Second period	Third period
Frequency (Hz)	Frequency (dB Hz)	$\Delta\varphi$ (°)	$\Delta\varphi$ (°)	$\Delta\varphi$ (°)
50	16.9897	4.4		
100	20	8.4		
500	26.9897	28.4		
1000	30	58.3		
2000	33.0103	127.9		
5000	36.9897	327.7		
10,000	40	295.7	623.4	
15,000	41.76091	262.3		885.7

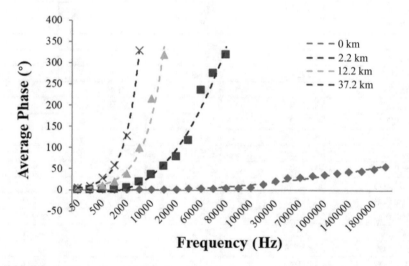

Fig. 5.25 Effect of frequency and fibre length on measured average phase

The above results prove this fact that the frequency of applied microwave signal is related with measured average phase of received optical power. By increasing the frequency from very low value to the high value, the average phase will grow up dramatically, whereas this trend for the longer length of fibre will change in very sharp increment. The below figure illustrates the phase change's trend for different length of fibres and frequencies (Fig. 5.25, Tables 5.11, 5.12, 5.13, and 5.14).

To calculate the dispersion from measured average phase, Eq. (5.8) could be useful.

$$\text{Phase} = \frac{\text{Delay}}{\text{Source1Period}} \times 360^{\circ} \tag{5.8}$$

The above tables are showing the characterization dispersion results based on different fibre lengths (Figs. 5.26, 5.27, 5.28, and 5.29).

Table 5.11 Dispersion characterization for 0 km

Phase (°)	Frequency (Hz)	Source period (s)	Delay (s)	Length (km)	Wavelength (nm)	Dispersion (S/nm m)
0.5	50	0.02	2.77778E-05	0	1540	1.80375E-08
0.5	100	0.01	1.38889E-05	0	1540	9.01876E-09
0.7	500	0.002	3.88889E-06	0	1540	2.52525E-09
1	1000	0.001	2.77778E-06	0	1540	1.80375E-09
1.1	2000	0.0005	1.52778E-06	0	1540	9.92063E-10
1.2	5000	0.0002	6.66667E-07	0	1540	4.329E-10
1.4	10,000	0.0001	3.88889E-07	0	1540	2.52525E-10
1.7	15,000	6.66667E-05	3.14815E-07	0	1540	2.04425E-10
2.1	20,000	0.00005	2.91667E-07	0	1540	1.89394E-10
2.2	30,000	3.33333E-05	2.03704E-07	0	1540	1.32275E-10
3	60,000	1.66667E-05	1.38889E-07	0	1540	9.01876E-11
3.6	70,000	1.42857E-05	1.42857E-07	0	1540	9.27644E-11
4	80,000	0.0000125	1.38889E-07	0	1540	9.01876E-11
4.6	900,000	1.11111E-06	1.41975E-08	0	1540	9.21918E-12
5.7	100,000	0.00001	1.58333E-07	0	1540	1.02814E-10
12.4	200,000	0.000005	1.72222E-07	0	1540	1.11833E-10
21.6	300,000	3.33333E-06	0.0000002	0	1540	1.2987E-10
27.6	500,000	0.000002	1.53333E-07	0	1540	9.95671E-11
30.5	700,000	1.42857E-06	1.21032E-07	0	1540	7.8592E-11
33.79	800,000	0.00000125	1.17326E-07	0	1540	7.6186E-11
35.7	1,000,000	0.000001	9.91667E-08	0	1540	6.43939E-11
39.39	1,200,000	8.33333E-07	9.11806E-08	0	1540	5.92082E-11

(continued)

Table 5.11 (continued)

Phase (°)	Frequency (Hz)	Source period (s)	Delay (s)	Length (km)	Wavelength (nm)	Dispersion (S/nm m)
42.7	1,400,000	7.14286E-07	8.47222E-08	0	1540	5.50144E-11
45	1,600,000	0.000000625	7.8125E-08	0	1540	5.07305E-11
50.39	1,800,000	5.55556E-07	7.77623E-08	0	1540	5.0495E-11
56.6	2,000,000	0.0000005	7.86111E-08	0	1540	5.10462E-11

Table 5.12 Dispersion characterization for 2.2 km

Phase (°)	Frequency (Hz)	Source period (s)	Delay (s)	Length (km)	Wavelength (nm)	Dispersion (S/nm km)
0.5	50	0.02	2.77778E-05	2.2	1540	8.19887E-09
0.5	100	0.01	1.38889E-05	2.2	1540	4.09944E-09
0.6	500	0.002	3.33333E-06	2.2	1540	9.83865E-10
0.7	1000	0.001	1.94444E-06	2.2	1540	5.73921E-10
5	2000	0.0005	6.94444E-06	2.2	1540	2.04972E-09
17.3	5000	0.0002	9.61111E-06	2.2	1540	2.83681E-09
37.2	10,000	0.0001	1.03333E-05	2.2	1540	3.04998E-09
56.29	15,000	6.66667E-05	1.04241E-05	2.2	1540	3.07676E-09
78.3	20,000	0.00005	0.000010875	2.2	1540	3.20986E-09
116.2	30,000	3.33333E-05	1.07593E-05	2.2	1540	3.1757E-09
235	60,000	1.66667E-05	1.08796E-05	2.2	1540	3.21122E-09
275.3	70,000	1.42857E-05	1.09246E-05	2.2	1540	3.2245E-09
318	80,000	0.0000125	1.10417E-05	2.2	1540	3.25905E-09
354.9	900,000	1.11111E-06	1.09537E-06	2.2	1540	3.23309E-10

Table 5.13 Dispersion characterization for 12.2 km

Phase (°)	Frequency (Hz)	Source period (s)	Delay (s)	Length (km)	Wavelength (nm)	Dispersion (S/nm km)
4.3	50	0.02	0.000238889	12.2	1540	1.2715E-08
4.9	100	0.01	0.000136111	12.2	1540	7.24458E-09
10	500	0.002	5.55556E-05	12.2	1540	2.95697E-09
20.5	1000	0.001	5.69444E-05	12.2	1540	3.03089E-09
37.9	2000	0.0005	5.26389E-05	12.2	1540	2.80173E-09
99.5	5000	0.0002	5.52778E-05	12.2	1540	2.94219E-09
214.9	10,000	0.0001	5.96944E-05	12.2	1540	3.17726E-09
318.6	15,000	6.66667E-05	0.000059	12.2	1540	3.1403E-09

Table 5.14 Dispersion characterization for 37.2 km

Phase (°)	Frequency (Hz)	Source period (s)	Delay (s)	Length (km)	Wavelength (nm)	Dispersion (S/ nm km)
4.4	50	0.02	0.000244444	37.2	1540	4.26694E-09
8.4	100	0.01	0.000233333	37.2	1540	4.07299E-09
28.4	500	0.002	0.000157778	37.2	1540	2.75412E-09
58.3	1000	0.001	0.000161944	37.2	1540	2.82685E-09
127.9	2000	0.0005	0.000177639	37.2	1540	3.1008E-09
327.7	5000	0.0002	0.000182056	37.2	1540	3.1779E-09

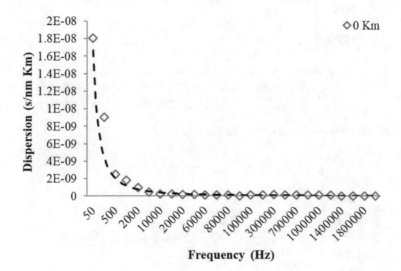

Fig. 5.26 Dispersion vs. frequency in 0 km

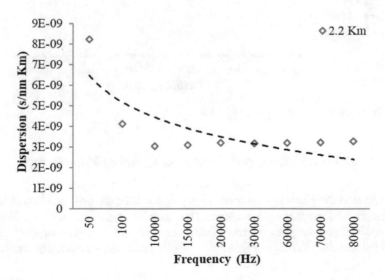

Fig. 5.27 Dispersion vs. frequency in 2.2 km

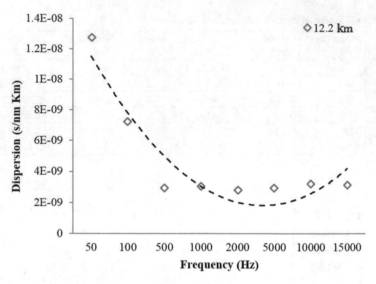

Fig. 5.28 Dispersion vs. frequency in 12.2 km

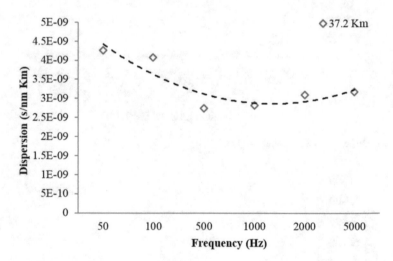

Fig. 5.29 Dispersion vs. frequency in 37.2 km

5.5   Analysis of Measured Phase vs. Variable Wavelength

In this section the effect of chromatic dispersion will be studied based on changing wavelength and measuring the phase. This method is also known as differential phase shift. Since the phase of modulated electrical signal is measured at intervals across the wavelength range of interest, at any two adjacent wavelengths, the change in group delay would be measured in picosecond (Eq. 5.9). Corresponding to the wavelength interval $\Delta\lambda$ is

$$\Delta T_\lambda = -\frac{\varphi_{\left(\lambda+\frac{\Delta\lambda}{2}\right)} - \varphi_{\left(\lambda-\frac{\Delta\lambda}{2}\right)}}{360 f_m} \times 10^{12} \tag{5.9}$$

where λ is the centre of wavelength interval, f_m is the modulation frequency in Hz and φ is the phase of recovered modulation signal. The delay in terms of measured phase and wavelength is showing by ΔT_λ parameter.

The following table and figure provide the specifications and the results of phase measurements based on change in wavelength (Fig. 5.30, Tables 5.15 and 5.16).

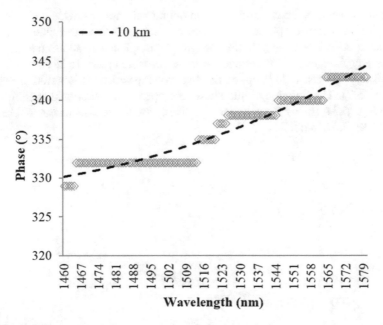

Fig. 5.30 Relative measured phase vs. wavelength for 10 km fibre

Table 5.15 Operational condition specifications for the phase measurement based on number of wavelength

EOM	792000280
Serial No	7F6e1900
V_b	8.5 V
V_p	1.1 V
Frequency	50 Hz

Table 5.16 Operation conditions to measure the phase vs. different V_b

Wavelength (nm)	1540
Frequency (Hz)	50
V_b (V)	2
Length (km)	0

By using differential phase shift method, the value of chromatic dispersion at the selected wavelength can be measured. Any changes in group delay across small wavelength interval can be applied to Eq. (5.10), and the average dispersion over the wavelength interval would be measured.

$$D_{\lambda_i} = \frac{\Delta\varphi_{\lambda_i} - \Delta\varphi_{\lambda_i}'}{360 f_m L \Delta\lambda} \times 10^{12} \qquad (5.10)$$

5.6 Analysis of the Effect of V_b on Measured Phase

In previous sections, the relation between output and input of EOM had been studied. Since the parameter phase is very sensitive due to its dependency to many other parameters such as wavelength, fibre length and frequency. In addition to the above factors, any changes on bias voltage cause the measured phase to be changed due to dependency of phase to the applied DC bias voltage which is discussed in Sect. 5.2 and Eq. (5.7). The following table shows the operation conditional specifications, while Fig. 5.31 shows how under the fixed condition, the phase is changing (Figs. 5.32, 5.33, and 5.34).

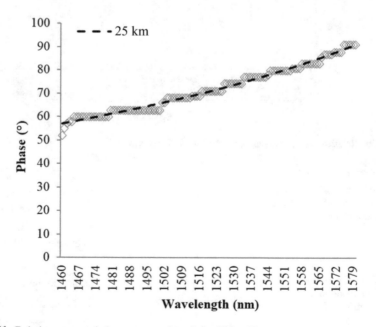

Fig. 5.31 Relative measured phase vs. wavelength for 25 km fibre

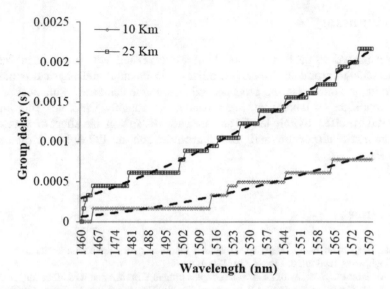

Fig. 5.32 Group delay measurement vs. wavelength at different fibre length

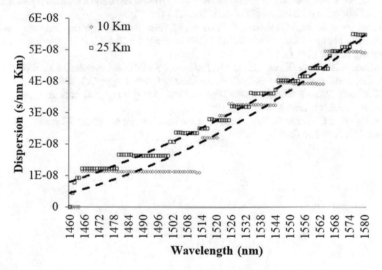

Fig. 5.33 Dispersion measurement vs. wavelength at different fibre length

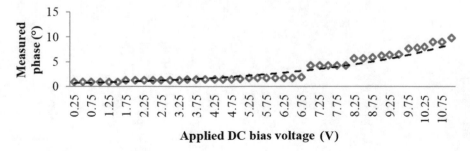

Fig. 5.34 Measured phase vs. V_b in EOM characterization

5.7   Summary

This chapter offers the final measurement results related with the setups in Chap. 4. In measuring chromatic dispersion, other environmental and physical conditions would impact on measuring progress and change the achieved results, but at standard condition of laboratory, these results are roughly trustable. This chapter reviewed the effect of fibre length on chromatic dispersion, the effect of wavelength on chromatic dispersion and the characterization of EOM based on critical parameters.

References

1. R.L. Jungerman et al., High speed optical modulator for application in instrumentation. J. Lightwave Technol. **8**, 1363–1370 (1990)
2. R. Madabhushi, A study of an antenna coupled optical Y branch and its application, Doctorate thesis for Doctor of Electronics Engineering, Tohoku University, Sendai, Japan, (1989)
3. K. Kawano, T. Nozawa, M. Yanagibashi, H. Jumoji, Broad band and low driving power LiNbO$_3$ external optical modulators. NTT Rev. **1**, 103–113 (1989)
4. K. Kawano, T. Kitoh, H. Junmoji, T. Nozawa, M. Yanagibashi, New traveling wave electrode Mach-Zehnder optical modulator with 20 GHz bandwidth and 4.7 V driving voltage at 1.52 μm wavelength. Electron. Lett. **25**, 1382–1383 (1989)
5. M. Seino, N. Mekada, T. Namiki, and H. Nakajima, 33-GHz-cm broadband Ti: LiNbO3 Mach-Zehnder modulator, in *Proc. ECOC, Paper ThB22–5*. (1989), pp. 433–435
6. M. Rangaraj, M. Minakata, A new type of Ti: LiNbO$_3$ integrated optical Y branch. IEEE Photonic. Technol. Lett. **1**, 230–231 (1989)
7. M. Rangaraj, M. Minakata, S. Kawakami, Low loss integrated optical Y branch. J. Lightwave Technol. **7**, 753–758 (1989)

Chapter 6
Optical Fibre Dispersions and Future Contributions on Electro-optic Modulator System Optimizations

6.1 Conclusion

In short, this work starts with a brief introduction in Chap. 1 about all general types of dispersion in fibre medium. Estimation and prediction of limitations caused by dispersion are essential considerations in designing any optical system. Hence, in the subsequent chapters, an automated system has been designed for measurement and characterization of dispersion. Chapter 2 discusses about the fundamental theory of dispersion and also the cause factors behind it. Appropriated electrical and optical devices with various functions were assembled for dispersion characterization purpose, and their optimum operating conditions have bccn studied in Chap. 3. The design of the system and the interlinked communication and control among programmable devices are illustrated in Chap. 4. The employed algorithms for the purpose of controlling, measuring and indexing results are also described in Chap. 4. Final measurement results are discussed practically and theoretically in Chap. 5.

6.2 Recommendations

Practically, the major component in characterization dispersion is EOM. For developing the system to work in an optimized way, first the bias point must be determined. Some factors must be applied to put the EOM in optimized condition. The following contributions could be proposed as future contributions of this work in order to design a fully utilized automatic controlling and measuring system:

- Designing automatic PC in a way that can adjust the polarization proportional to the output of the EOM in order to achieve the highest optical intensity and highest stability in received optical signal might be proposed as future work.

I. S. Amiri, M. Ghasemi, *Design and Development of Optical Dispersion Characterization Systems*, SpringerBriefs in Electrical and Computer Engineering, https://doi.org/10.1007/978-3-030-10585-3_6

- The amount of injected power from EDFA is the part of compensation process to overcome the attenuation caused by EOM. Since the amount of amplification is related with the distance between transmitter and receiver, the algorithm is required to carry out it and also avoid the *laser diode* (LD) to enter to the saturation threshold.
- Fully utilized characterization dispersion system required the feedback between TLS, microwave source and DOSC in order to measure dispersion at different frequencies and wavelengths.

Index

© The Author(s), under exclusive license to Springer Nature Switzerland AG 2019   69
I. S. Amiri, M. Ghasemi, *Design and Development of Optical Dispersion
Characterization Systems*, SpringerBriefs in Electrical and Computer
Engineering, https://doi.org/10.1007/978-3-030-10585-3